Agrarian Crisis in India

Agrarian Crisis in India

edited by
D. NARASIMHA REDDY
SRIJIT MISHRA

OXFORD
UNIVERSITY PRESS

OXFORD
UNIVERSITY PRESS

Oxford University Press is a department of the University of Oxford.
It furthers the University's objective of excellence in research, scholarship,
and education by publishing worldwide. Oxford is a registered trademark of
Oxford University Press in the UK and in certain other countries

Published in India by
Oxford University Press
22 Workspace, 2nd Floor, 1/22 Asaf Ali Road, New Delhi 110002, India

First Edition published in 2009
Oxford India Paperbacks 2010
17th impression 2023

ISBN-13 (print edition): 978-0-19-806909-6
ISBN-10 (print edition): 0-19-806909-x

ISBN-13 (eBook): 978-0-19-908830-0
ISBN-10 (eBook): 0-19-908830-6

Typeset in Minion pro 10.5/12.7 by Jojy Philip
Printed in India at Replika Press Pvt. Ltd., Haryana 131 028

Contents

Tables, Figures, Annexures, and Boxes

FIGURES

Foreword

There has been a distinct slowdown in agricultural growth since the mid-1990s which has adversely impacted the livelihood base of the farming community at large. The slowdown has occurred in all the sub-sectors of agriculture, including livestock and horticulture which were the main drivers of agricultural growth in the immediate past. A large number of proximate and structural factors have contributed to the decline of agriculture. The foremost among them is the reduced developmental role of state in investment in irrigation, flood control, research, extension, and institution building in the context of liberalizing agriculture. This has not only affected the externalities that induce private investment and provide an environment conducive for innovation by the farmers, but also the generation of agricultural technology in public institutions as well as extension. Some of the agrarian institutions started decaying as they could not adapt to the new challenges brought about by changes in the policy regime. Moreover, new institutions have not been put in place to mitigate the risks generated by the changes in the policy regime. Without adequate intervention, the pace of change was allowed to have its full sway on the livelihood of the farmers, especially smallholders. The institutional retardation has begun to happen just when the farming community lost its capacity to generate self-equilibrating response to macroeconomic changes. The liberalization of agricultural trade has exposed commercial agriculture to the volatility in the world commodity markets. Furthermore, when agricultural prices in the world market were declining in the later half of the 1990s and the first half of the 2000s, India dismantled its quantitative restrictions and slashed the tariff rates. The recent rise in the international cereal prices may have provided price incentives to the farmers but has simultaneously hurt the poor the most. The burden of falling international prices during 1996–2004 fell on the farmers but gains from the subsequent rise in food prices might have accrued disproportionately more to the middlemen between the direct producer and consumer. The desirable goal of agricultural growth with stability has become more distant, since the existing instruments are too inadequate to mitigate the risks affecting the farmers.

The present book edited by D. Narasimha Reddy and Srijit Mishra contains a set of contributions dealing with the agricultural distress. The essays put together provide a very comprehensive analysis of the macroeconomic and micro-level issues associated with the agricultural distress. While the focus of the book is on the last one decade of agrarian crisis, the analysis is sensitive to the historical forces that underlie the crisis. It discusses the factors contributing to the crisis: the degradation of the environment, dwindling of land holding size, plateauing of the yields from the present farm technology, and, most important, withdrawal of the state support.

In the post-reform period, high gross domestic product (GDP) growth accompanied by low agricultural growth has brought about a rapid shift in the sectoral distribution of GDP, which is skewed against agriculture (Reddy and Mishra). The declining share of agriculture is a tendency driven by forces inherent in the development process; however, the paradox is that there is no commensurate decline in the share of agriculture in the country's labour force. The book documents that rapid growth of GDP has not created enough opportunities for rural labour force to move out of agriculture into the rapidly growing sectors of the economy. Low growth of farm productivity, low agricultural prices, slowdown of demand for agricultural products due to stagnation of per capita food consumption during 1993–2005, and inadequate employment opportunities outside agriculture are the proximate causes that underlie the current slow pace of agricultural transformation.

The book makes an earnest effort to bring together the supply-side factors. There is a consensus that growth of irrigation has slowed down owing to a decline in public investment in irrigation infrastructure. The addition to irrigated area was only 0.8 million hectares per annum during the 1980s and 1990s. This is in contrast to annual addition of 2.5 million hectares to irrigated area during the Green Revolution. Public expenditure on research and education continued to remain low at around 0.5 per cent of GDP (on an average, developing countries spend 0.7 per cent of GDP and developed countries spend 2 to 3 per cent of GDP). It is important to recognize that the research requirements are high in India in view of substantial variations in agro-climatic conditions that warrant region-specific and crop-specific technologies compatible with the endowments of the farmers. Needless to say, efforts in this direction are grossly inadequate (Suresh Pal). It is worrying that no major technological innovations which could make decisive impact on agricultural productivity are in the offing, particularly in the dryland agriculture where the distress is more. What is worse, in many states the agricultural extension system has virtually collapsed (Reddy and Mishra).

A major area of concern highlighted by S.L. Shetty is the sluggish growth of institutional credit. The share of agriculture in the institutional credit, at about 10–11 per cent, was far below the stipulated target of 18 per cent. Half of the farmers had no access to institutional finance in 2003. Moreover, the institutional agencies accounted for 57.7 per cent of the outstanding loan amount of farmers, followed by moneylenders (25.7 per cent), and traders (5.2 per cent). These data suggest heavy dependency of farmers on informal sources of finance. The picture is worse for small and marginal farmers. Interest rates charged by informal sources are not affordable, given the low productivity levels in agriculture. For instance, about 40 per cent of the cash debt from informal sources outstanding in 2003 was at the interest rate of 30 per cent or more. In contrast, the interest rate was less than 20 per cent for 99 per cent of the debt from institutional sources. Recent decisions to reduce interest rate to 7 per cent and double the credit flow from institutional sources are welcome steps. However, some of the credit-related measures such as redefinition of priority sector and doubling of credit have the tendency to promote bimodal growth within rural areas. For instance, term credit by the scheduled commercial banks in the recent years, which is a major driver of investment in agriculture, supported more capital-intensive sector such as high-tech, storage, new markets, and farm mechanization, and provided less support for minor irrigation, land development, fisheries, etc. While investment in agricultural allied activities and infrastructure is necessary to promote value addition, it is equally important to increase the investment in traditional activities of agriculture for improving the incomes of the smallholders. Further, the share of large farmers (with land holdings above 5 acres) in the commercial banks' outstanding credit increased from 46 per cent in 1991–2 to 52 per cent in 2003–4.

The supply side factors in the recent period have exerted downward pressure on per hectare farm business income, and furthermore, the farm business income continued to experience sharp year-to-year variations (Reddy and Mishra). These unfavorable factors have put at risk the livelihood of about 60 per cent of the population. The recent spate of farmers' suicides is a symptom of the agrarian distress caused by declining profitability of agriculture and increasing risks due to growing commercialization of agriculture. The case studies on farmer suicides included in the book provide insights into their underlying causes. They suggest that suicides are symptomatic of a deep-rooted crisis in agriculture. It is argued that farmers' distress is a serious growth-retarding factor in the long run since a sizeable proportion of farmers lost capacity to withstand shocks; what is worse is that they are prepared to leave agriculture if other opportunities are available (V.M. Rao).

The agricultural distress, as expected, had a greater impact on the smallholders with limited resources and the farming community inhabiting the rain-fed regions and tribal areas. Reddy and Mishra pose a pointed question—whether small and marginal farming is sustainable without public infrastructure support and comprehensive social security covering health, education, employment, and old age support? Further, the public funding for research and education is low in dryland states with harsh environments. The increasing capital intensity of new technology and institutional bias in the delivery of technology restricts access to technology for small and marginal farmers (Suresh Pal).

These essays suggest that revival of agriculture requires tackling of two dimensions of the agricultural distress—agricultural development crisis (reflected in low growth, declining profitability of agriculture) and agrarian crisis (reflected in growing landlessness and casualizaton of labour in agriculture, unchecked proliferation of small and marginal holdings, fragmentation of landholding, and widening gap between rural and urban areas). The declining role of agriculture is inevitable in the development process. However, the transition needs to be managed by substantially raising farm productivity and, at the same time, rapidly shifting the labour force from agriculture to growing sectors of the economy. Given the poor quality of public institutions, absence of institutions for managing risks, increasing pressure on land and water resources, and adverse macro environment, the prospects of overcoming farmers' distress is more challenging than is realized. There is less appreciation in the policymaking circles of the fact that tackling the agrarian crisis is far more difficult than reviving agricultural growth. The essays included in this book, besides analysing the causes for the current crisis in agriculture in different parts of the country, emphasize the need for improved public investment in agricultural infrastructure and the social overheads like quality education and health facilities in rural areas, and augmented expenditure on research and development as well as extension that would involve a combination of modern technology with traditional biotic practices. Improving access to adequate institutional finance and marketing facilities, especially to small and marginal farmers, and ensuring proper regulatory system to provide quality seeds and other inputs at reasonable prices are some of the measures suggested for the revival of agricultural growth that would ensure improved conditions of living for farmers.

R. RADHAKRISHNA
Hon. Professor of Economics
Centre for Economic and Social Studies
Hyderabad

Preface

For over a decade, while the Indian economy has been experiencing unprecedented rate of high growth, agriculture has been passing through a phase of deceleration in growth, and there has been widespread distress manifesting in suicides of farmers. There is a wide recognition that the crisis in agriculture is a result of deep seated malady and that the suicides are only symptoms. The crisis assumes different intensities under different conditions. For instance, it is survival crisis in dry regions like northern Karnataka or southern Andhra Pradesh. It is sustainability crisis in prosperous regions like Punjab or plantations of Kerala. The economic reforms initiated in early 1990s not only failed to help agricultural growth but have actually aggravated the situation. Though foodgrains did register a rate of growth less than population growth during the 1990s, it is not merely crisis of deficits in production. In general, returns in agriculture did dwindle but it is the small farmers in non-command areas attempting to be upwardly mobile through heavy borrowing for investment who were trapped in serious crisis in the absence of adequate and appropriate state support services and volatile markets. To capture the complexity, it is useful to distinguish the two faces of the crisis, viz., 'agrarian crisis' and 'agricultural crisis'. Such a framework will enable to differentiate dimensions of the crisis in terms of substance and adopt appropriate methods of analysis. 'Agrarian crisis' is structural and institutional in nature as could be seen in growing marginalization and failure of support systems especially as a part of the reforms agenda because of the shift in institutional emphasis from state to market. Methodologically, the analysis of the impact of structural and institutional changes requires household and village level studies with particular attention to the processes of change. 'Agricultural crisis', on the other hand, may be seen in terms of performance of production in relation to the problems associated with access and use of inputs and realization of returns. Methodologically, a fair amount of time series and large-scale survey data are available on the behaviour of input costs, output prices, resource degradation, inappropriate technology, and technological fatigue. What is more important is that the complexity of the present crisis needs to be seen in its inter-relatedness between agrarian crisis and agricultural crisis and analysed as to how one reinforces the other.

The present book is a modest attempt towards explaining the crisis in all its dimensions. It consists of essays that address the macro dimensions as well as micro-level manifestations. The methods used in these essays range from models of explanation based on aggregate data from large national surveys to descriptive analysis based on case studies of farmer households with incidence of suicides. The book is divided into two parts. The first part, 'Macro Dimensions of Agrarian Crisis' includes four contributions. The second part, 'Farmers' Distress: A Few States in Focus', includes six contributions, five of which deal with farmers' distress in five different states, and one synthesizes the lessons from the five case studies. The first chapter, 'Agriculture in the Reforms Regime' provides an overview of the dimensions of crisis in agriculture, particularly since the initiation of economic reforms in the early 1990s. Structurally shrinking share of agriculture in the national product with continued concentration of the workforce, and rapid marginalization of holdings with growing productivity distance to non-agriculture are analysed. Increasing stress on resources, especially irrigation, environmental degradation, and technological fatigue are other factors identified with productivity plateau. The impact of reforms including trade liberalization, decline in public investment in agriculture, shrinking share of formal institutional credit, rising input prices, and volatility in commodity prices are identified as factors contributing to agrarian distress. The chapter reviews the evidence from other studies on farmers' suicides and argues that in the context of small farmer agriculture, there is need for improvement in state support systems, including social sector investment, to reduce the compounded risks of adjustments in agriculture towards improved productivity and livelihoods. The second chapter, 'Capital Formation in Indian Agriculture: National and State Level Analysis' analyses the trends in public and private capital formation in agriculture during the past quarter century. It brings out the drastic decline in the growth of public investment in agriculture in the 1980s and virtual stagnation beginning with the early 1990s. Attention is also drawn to the negative impact of growing subsidies on public investment in agriculture. However, it is shown that the combined public outlay on capital formation and subsidies in agriculture as a share of national product declined during the 1990s and it is attributed to anti-agricultural bias of the reforms regime. By using capital expenditure on agriculture and related activities as an indicator of public capital formation at the state level, it is found that there was weakening of public investment in agriculture. The chapter emphasizes the importance of public investment in reviving agricultural growth.

The third chapter, 'Agricultural Credit and Indebtedness: Ground Realities and Policy Perspectives', brings out the trends in the growth of institutional credit to agriculture. It is shown that during the years following bank nationalization, there was rapid increase in the institutional credit to agriculture at a rate much faster than the growth of agricultural Gross Domestic Product (GDP). However, the increasing credit needs of agriculture because of commercialization were not fully met, particularly in the 1990s, which resulted in increasing dependence of especially small and marginal farmers on expensive informal sources. There was decline in the credit flow to agriculture during the 1990s because of financial sector reforms. Efforts towards better flow of institutional credit to agriculture were revived in late 1990s but attention needs to be paid to better credit delivery and effective monitoring. The fourth chapter, 'Managing Vulnerability of Indian Agriculture: Implications for Research and Development', shows that there has been steady increase in public funding for research and education since 1981, and unlike in other sectors, public investment in research increased sharply in late 1990s. Despite that, in 2000–1, public funding for agricultural research was only 0.31 per cent of agricultural GDP and it is just half of the average for developing countries. Further, it is shown that public effort in developing hybrids is limited, leaving the market to the private operators, resulting in very high cost of seed replacement to farmers. The essay emphasizes the need for an effective regulatory mechanism for ensuring supply of quality seed at affordable prices to farmers.

The second part of the book begins with 'Farmers' Distress in a Modernizing Agriculture—The Tragedy of the Upwardly Mobile: An Overview', which characterizes farming communities into a three-tier pyramid formed as a result of the policies pursued over the years. At the wide base of the pyramid is the myriad mass of poverty-stricken farmers, in the middle are the 'upwardly mobile farmers'—meaning those struggling to move up the productivity level but often sinking than swimming—and at the top is a small section lobbying for subsidies, seeking rents or unproductive profits rather than use their entrepreneurial potential for agricultural growth. The chapter observes that it is activating and providing state support to the middle-upwardly-mobile stratum which could make growth inclusive even to those who are at the bottom of the pyramid. Turning to the case studies of suicides in five states, a pen picture of a typical farmer struggling to move up in the drought prone areas is constructed, and critical interventions in terms of credit, extension services, and institutional structure are suggested. The five chapters that follow are case studies of states in severe agrarian distress. Chapter 6, 'Agrarian Distress and Farmers' Suicides in Maharashtra', begins with the structural characteristics of the state. In spite of being one of the leading industrial states, there has been continued dependence of majority of the workforce on agriculture, even as its share in state domestic product has drastically declined. Then it turns to trends in public investment, credit, and related issues in agriculture. The main part of the chapter deals with suicides in western Vidarbha, a rain-fed region of the state. The study shows that farmers face multiple risks like uncertain weather, volatile markets, lack of proper technology, spurious inputs, and inadequate institutional credit, but indebtedness turns out to be a critical factor triggering suicides. The analysis of the Union government's intervention package shows that though it is comprehensive, it suffers from several deficiencies in design and implementation. The seventh chapter, 'Farmers' Suicides and Unfolding Agrarian Crisis in Andhra Pradesh', begins with an account of farmers' suicides in the state beginning with 1996. A survey of several studies shows that farmers' distress in the state is due to growing fragility of resources, especially the groundwater resources, unsustainable and inappropriate cropping patterns especially in resource poor areas. Field-level survey by the authors and secondary sources of information from cost of cultivation data show that the crisis is due to a number of factors of which the most important are: growing dependence on unstable groundwater resources, heavy household investment in groundwater by borrowing from informal sources at high rates of interest that draws farmers into debt trap, growing input costs, volatility of output prices, and declining farm incomes. The eighth chapter, 'Agrarian Transition and Farmers' Distress in Karnataka', draws attention to the fact that the state has one of the lowest proportions of irrigated area with the predominance of rain-fed agriculture. Under such circumstances, the narrow techno-centred green revolution strategy brought about inappropriate cropping patterns in many parts. The state is characterized by extreme regional disparities, and agricultural distress is acute in the northern dry regions with a very high incidence of suicides since 1997. Besides indebtedness, increasing stress on natural resources, frequent failure of monsoon and droughts leading to crop losses, lack of proper marketing facilities, declining extension services, and absence of counselling institutions are identified as the main factors precipitating the crisis.

The ninth chapter, 'Distress, Debt, and Suicides among Farmer Households: Findings from Village Studies in Kerala', is based on a survey of three villages in Wayanad and Idukki districts with plantation-based agriculture. Both the districts faced severe agrarian crisis since the year 2000 and witnessed several suicides of farmers. There was sharp decline in prices of plantation crops like pepper, coffee, cardamom, and tea. This was accompanied by decline in yields due to adverse weather conditions and pests. The policy changes of trade liberalization in general, free trade agreement with Sri Lanka in particular, and declining state support aggravated the situation. The tenth chapter, 'Agrarian Crisis in Punjab: High Indebtedness, Low Returns, and Farmers' Suicides', analyses the factors behind agrarian crisis in one of the most prosperous states in India. It shows that there has been a significant deceleration in agricultural growth since the early 1990s. Based on the cost of cultivation data, it is shown that profitability in agriculture has been on the decline due to cost of inputs increasing faster than output prices, combined with stagnation or even decline in yields. Rising cost of capital equipment like tractors, substantial idle capacity in capital equipment, and increasing cost of irrigation due to steep fall in the water table are other factors that eroded incomes from agriculture. Though there are numerous factors that lead farmers to suicides, economic stress due to indebtedness is the trigger.

* * *

The contributions in the volume together provide a fairly comprehensive picture of the state of agriculture in India at the turn of the twenty-first century. Yet, there are important lacunae like inadequate attention to dry regions, regional diversities and disparities, potential of structural changes that improve access to land, institutional agencies like small farmer organizations, emerging alternative technologies, and of course, non-farm sector linkages. Much of this would hopefully be part of the debates on the future of agriculture in India. In putting together this volume, we are indebted to a large number of people who have sustained our efforts in various ways. The credit for initiating, nurturing, and providing logistic and financial support to this project goes entirely to R. Radhakrishna, the former Director of the Indira Gandhi Institute of Development Research, Mumbai. We are also grateful to him for writing the foreword to this book. We thank V.M. Rao who was also part of the initiation, and sustained us by providing his moral and intellectual support all through. S.R. Hashim, S.L. Shetty, Amit Bhaduri, G.S. Bhalla, Shiela Bhalla, V.S. Vyas, and Manoj Panda helped us in shaping this project by providing their comments and encouragement at various stages. We are highly indebted to them and look forward to their active participation in the debate that the book is intended to generate. The contributors to this volume are all busy scholars with tight schedule of work of their own and of their respective institutions. In spite of that, they have cooperated wonderfully well by participating in the authors' workshop, by reviewing some of the essays and by admirably meeting the deadlines of revisions. We thank them, and only wish we could have had resources to adequately compensate their intellectual input. Late P.D. Jeromi, B.N. Kulkarni, B.B. Mohanty, R.M. Mohan Rao, G.V.

Ramanjaneyulu, and Krishna Vatsa also helped us in reading and commenting on the earlier drafts of some of the papers for which we are grateful to them. We also put on record that all the papers were subjected to anonymous reviewing by two peers, as would be the case for submissions in professional journals. All through the project, the administrative staff of IGIDR and of the Guest House have been of great help, and we are thankful to all of them. The staff at Oxford University Press, New Delhi, were of constant help through their timely reminders and faster completion of the referring process, for which we are grateful to them. And it is our pleasure to thank Sarthak, Srinivas, Yogesh, and specially Satish who patiently suffered our untimely demands on their word processing skills.

Abbreviations

ADRT	Agricultural Development and Rural Transformation
AFDR	Association of Federal Democratic Rights
AgGDP	Agricultural Gross Domestic Product
AICRP	All India Coordinated Research Programme
AIDIS	All India Debt and Investment Survey
AOA	Agreement on Agriculture
APERP	Andhra Pradesh Economic Restructuring Project
APRS	Andhra Pradesh Rythu Sangam
AWARE	Action for Welfare and Awakening in Rural Environment
BBC	British Broadcasting Corporation
BC	Backward Communities
BPL	Below Poverty Line
BSR	Basic Statistical Returns
CACP	Commission for Agricultural Costs and Prices
CCB	Cooperative Central Cooperative Bank
CCDI	Comprehensive Composite Development Indices
CCI	Cotton Corporation of India
CCS	Cost of Cultivation Scheme
CDR	Crude Death Rate
CES	Centre for Environmental Studies

CGCIL	Credit Guarantee Corporation of India Limited
CGS	Credit Guarantee Scheme
CGTSI	Credit Guarantee Fund Trust for Small-scale Industries
CPI	Consumer Price Index
CPIAL	Consumer Price Index for Agricultural Labour
CSO	Central Statistical Organisation
FBI	Farm Business Income
FGD	Focus Group Discussion
GCF	Gross Capital Formation
GDP	Gross Domestic Product
GFCF	Gross Fixed Capital Formation
GLC	General Lines of Credit
GM	Genetically Modified
GSDP	Gross State Domestic Product
GSP	Generalized System of Preferences
ha	hectare/s
HH	Household
HPCRRI	High Power Committee for Redressal of Regional Imbalances
HYV	High Yielding Variety
IBH	India Book House
ICAR	Indian Council of Agricultural Research
IGIDR	Indira Gandhi Institute of Development Research
IGP	Indo-Gangetic Plain
IMF	International Monetary Fund
IPR	Intellectual Property Rights
ISEC	Institute for Social and Economic Change
KCC	Kisan Credit Card
KVK	Krishi Vigyan Kendra
MARKFED	Marketing Federation Limited
MCPS	Monopoly Cotton Procurement Scheme
MEGS	Maharashtra Employment Guarantee Scheme
MNC	Multinational Companies

MOF	Ministry of Finance
MOSPI	Ministry of Statistics and Programme Implementation
MoU	Memorandum of Understanding
MPCI	Monthly Per Capita Income
MSP	Minimum Support Price
NABARD	National Bank for Agriculture and Rural Development
NAS	National Accounts Statistics
NCAER	National Council of Applied Economic Research
NCAP	National Centre for Agricultural Economics and Policy Research
NCCR	National Centre of Competence in Research
NCRB	National Crime Records Bureau
NGO	Non-governmental Organization
NHB	National Housing Bank
NIA	Net Irrigated Area
NNP	Net National Product
NPAs	Non-performing Assets
NPK	Nitrogen, Phosphorus, and Potassium
NRC (LTO)	National Rural Credit (Long Term Operations)
NREGP	National Rural Employment Guarantee Programme
NRSA	National Remote Sensing Agency
NSA	Net Sown Area
NSDP	Net State Domestic Product
NSS	National Sample Survey
NSSO	National Sample Survey Organisation
OBC	Other Backward Classes
OGL	Open General Licensing
OPVs	Open Pollinated Varieties
PACS	Primary Agricultural Cooperative Credit Society
PAU	Punjab Agricultural University
PCI	Per Capita Income

PLR	Prime Lending Rates
PPV&FR	Protection of Plant Varieties and Farmers' Rights
PRIs	Panchayati Raj Institutionss
PRO	Pragatepar Raitha Okkuta
PSFC	Punjab State Farmers Commission
PSASMU	Poverty and Social Monitoring Unit
QR	Quantitative Restrictions
qtl	quintal
R&D	Research and Development
R&E	Research and Education
RBI	Reserve Bank of India
RCE	Real Capital Expenditure
RFAS	Rural Finance Access Survey
RRB	Regional Rural Bank
Rs	Rupees
RSC	Ryaithu Sahaya Committee
SADB	State Agricultural Development Bank
SAP	Structural Adjustment Programme
SAS	Situation Assessment Survey
SAU	State Agricultural Universities
SCB	Scheduled Commercial Bank
SERP	Society for Elimination of Rural Poverty
SIDBI	Small Industries Development Bank of India
SME	Small And Medium-Enterprises
SMR	Suicide Mortality Rate
SSI	Small Scale Industries
TE	Triennium Ending
TFP	Total Factor Productivity
UPSS	Usual Principal and Subsidiary Status
VAO	Village Agricultural Officer
WTO	World Trade Organization
WUA	Water Users' Associations

Macro Dimensions of
Agrarian Crisis

1

Agriculture in the Reforms Regime[*]

D. Narasimha Reddy and Srijit Mishra

Introduction

Economic reforms initiated in 1991 have set the Indian economy on a high growth path, making it one of the leading emerging market economies of the world. The institutional shift from planned but slow growth process to market-driven rapid growth has, however, been accompanied by rising income and regional disparities, deceleration in the reduction of poverty, and rural distress embedded in agrarian crisis.

This chapter addresses the nature and causes of the unfolding agrarian crisis and rural distress and is divided into five sections. The first section draws attention to the fact that the Indian economy is still predominantly rural, with slow urbanization but growing rural–urban disparities in income and levels of living. Agriculture continues to be the most important source of livelihood in the countryside. It accounts for a disproportionately high share in the total workforce, but with a rapidly declining share in the national product. The first section discusses the structural changes in employment across the sectors and within rural areas. It also analyses the increasing marginalization of holdings and the emergence of small-marginal holdings as a numerically dominant group in Indian agriculture. The second section deals with the growing resource stress experienced by agriculture. The third section examines the deceleration of agricultural growth in the post-reform period. The fourth section draws attention to the political economy of the technological transformation in

Indian agriculture. The fifth section analyses the nature of economic reforms with specific reference to Indian agriculture and their impact on farming and the farming community. The sixth section brings out the broad contours of agrarian crisis that manifests in the form of a series of suicides. The last section deals with the sustainability of farming in the context of reforms and concludes with some reflections on a possible way out.

STRUCTURAL CHANGES IN THE INDIAN AGRICULTURE

Demographic Pressure on Agriculture

Even at the beginning of the twenty-first century, India has continued to be rural and agricultural in terms of livelihood activities of people. In 1999–2000, 72 per cent of the population and 76 per cent of the workforce in India were rural, accounting for about one-fifth of the national income. Within rural areas, there has, in a way, been excessive dependence on agriculture. Table 1.1 shows that even during the period of economic reforms of the 1990s and the much-lauded high economic growth, there was no substantial increase in the share of the rural non-farm sector. The employment status of rural labour tends towards relatively more insecure casual labour, while self-employment and regular employment show a declining share.

Table 1.1
Sectoral Share and Employment Status of Rural Workforce

(per cent)

Rural Employment	1983	1987–8	1993–4	1999–2000	2004–5
Agriculture	81.49	77.46	78.39	76.16	70.08
Non-agriculture	18.51	22.54	21.61	23.84	29.92
Status of rural workforce					
Self-employed	61.37	59.50	57.96	55.76	60.2
Hired–regular	7.15	7.79	6.45	6.83	7.1
Hired–casual	31.49	32.72	35.59	37.41	32.8

Source: Bhalla (2005); National Sample Survey Organisation (NSSO), *Employment and Unemployment Situation in India*, various rounds.

Table 1.2
Share of Agriculture in GDP and Employment

(per cent)

Year	Share of Agriculture in GDP at 1999–2000 Prices	Share of Agriculture in Employment (UPSS)
1972–3	41.0	73.9
1993–4	30.0	63.9
1999–2000	25.0	60.2
2004–5	20.2	56.5

Source: Central Statistical Organisation (CSO); National Accounts Statistics, various years; and NSSO, Employment and Unemployment Situation in India, various rounds.

An important structural feature of the Indian economy has been a continuous decline in the share of agriculture in total gross domestic product (GDP), but very slow diversification of the workforce away from agriculture. While the share of agriculture in GDP declined from 41 per cent in 1972–3 to 20 per cent in 2004–5, the share of employment in agriculture (UPSS) declined from 74 per cent in 1972–3 to only 57 per cent by 2004–5 (Table 1.2).

The slow growth of employment in the formal sector, and agriculture as employment of last resort, results in structural rigidity that inhibits the shift of labour force from agriculture to non-agriculture, contributing to a very large and increasing gap in the per worker earnings in the agriculture and non-agricultural sectors. This has resulted in a persistent decline in the relative per worker productivity and income of agricultural workers vis-à-vis non-agricultural workers. With 56.5 per cent of the national workforce producing only about one-fifth of the GDP, the relative productivity of workers in agriculture has declined from 28.7 per cent of non-agricultural productivity in 1972–3 to 19.9 per cent in 2004–5. This should be a matter of grave concern for policy makers in India. The concentration of workforce in the agricultural sector is much higher in rural areas, indicating that the rural economy continues to remain by and large an undiversified economy, primarily dependent on agriculture. This is true despite relatively lower growth of employment in agriculture during the recent decades (Table 1.3).

Table 1.3

Distribution of Usually Working (UPSS) by Broad Group of Industry:
Rural India

(per cent)

Year	Male			Female		
	Primary	Secondary	Tertiary	Primary	Secondary	Tertiary
1983	77.5	10.0	12.2	87.5	7.4	4.8
1987–8	74.5	12.1	13.4	84.7	10.0	5.3
1993–4	74.1	11.2	14.7	86.2	8.3	5.5
1999–2000	71.4	12.6	16.1	85.4	9.0	5.8
2004–5	66.5	15.5	18.0	83.3	10.2	6.6

Note: Workers denote Usual Principal and Subsidiary Status.
Source: NSSO, *Household Consumer Expenditure and Employment Situation in India*, various rounds.

Increasing Marginalization

Lack of employment diversification has resulted in a concentration of workforce in the agricultural sector. Increasing the workforce on non-expanding cultivable land leads to an increased number of holdings and decreased size of holdings. Between 1960–1 and 2003, the number of holdings increased from 51 million to 101 million and the area operated declined from 133 million hectares to 108 million hectares (Table 1.4). Consequently, the size of operational holdings declined from 2.63 hectares in 1960–1 to only 1.06 hectares by 2003. Even assuming that 2003 was a drought year, Table 1.4 shows that the operated area since the 1970s did not exceed about 125 million hectares, which would still leave the average size of holding at not more than 1.23 hectares, an area that is too small to provide adequate livelihood.

As discussed above, the increasing demographic pressure on land has resulted in undue stress on land resources, and reduced the size of holdings to uneconomic levels. Added to this is the fact that despite land reforms, land continues to be distributed in a very skewed manner. Table 1.5 shows clearly that over the years, there has been increasing concentration of marginal and small farmers and that the proportion of marginal farmers operating less than one hectare of land is increasing at a very fast rate. The process of marginalization has resulted in wide variations in the income and living standards of various categories of cultivators and landless labourers.

Table 1.4
Key Characteristics of Operational Holdings in India

Characteristics	1960–1	1970–1	1981–2	1991–2	2003
	(17th)*	(26th)*	(37th)*	(48th)*	(59th)*
Number of operational holdings (millions)	50.77	57.07	71.04	93.45	101.27
Percentage increase	–	12.4	24.5	31.5	8.4
Area operated (million hectares)	133.48	125.68	118.57	125.10	107.65
Average area operated (hectares)	2.63	2.20	1.67	1.34	1.06

Note: * Indicates NSS rounds.
Source: NSSO, Some Aspects of Operational Land Holdings in India, various rounds.

Table 1.5
Changes in the Size Distribution of Operational Holdings and Operated Area: 1960–1 to 2002–3

(per cent)

Size class	Percentage of operational holdings					Percentage of operational area				
	1960–1	1970–1	1981–2	1991–2	2003	1960–1	1970–1	1981–2	1991–2	2003
	17th	26th	37th	48th	59th	17th	26th	37th	48th	59th
Marginal	39.1	45.8	56.0	62.8	71.0	6.9	9.2	11.5	15.6	22.6
Small	22.6	22.4	19.3	17.8	16.6	12.3	14.8	16.6	18.7	20.9
Semi-medium	19.8	17.7	14.2	12.0	9.2	20.7	22.6	23.6	24.1	22.5
Medium	14.0	11.1	8.6	6.1	4.3	31.2	30.5	30.1	26.4	22.2
Large	4.5	3.1	1.9	1.3	0.8	29.0	23.0	18.2	15.2	11.8
All sizes	100.0	100.0	100.0	100.0	100.0	100.0	100.0	100.0	100.0	100.0

Source: NSSO, Some Aspects of Operational Land Holdings in India, various rounds.

RESOURCE STRESS IN INDIAN AGRICULTURE

Increasing Stress on Irrigation Resources

Added to the problem of smallholdings is the unequal availability of irrigation across the country, and increasing stress on available irrigation resources. India is not very rich in irrigation resources—with 16 per cent of the world's population, it is endowed with only 4 per cent of the total available freshwater. Furthermore, the regional distribution of available water resources, including rainfall, is highly uneven. With demand for drinking water and other needs associated with rapid urbanization and industrialization increasing at a very rapid rate, there are going to be serious challenges for adequate availability of water for irrigation. The growing and competing demands for water are bound to emerge as a serious challenge to water resource management.

The recent trends in irrigation show the distortion in the development and utilization of water resources for agricultural purposes (Table 1.6). It is well known that one of the major areas of public investment in post-Independence India has been the investment on major and medium irrigation projects, which contributed to a substantial expansion of areas irrigated. In the post-reform period, however, there has been a net decline in the area irrigated under canals. The Plan era also showed neglect of minor surface irrigation sources such as tanks, leading to decay and disuse of these waterbodies. The pace of decline in the area irrigated by tanks accelerated to –3.15 beginning with 1990–1. The only source that has been continuously on the increase, which by 2003–4 accounted for almost two-thirds of the net irrigated area in the country, is groundwater exploitation through wells and borewells, though the rate of growth of even this resource is slowing down because of increasing risks and limits to the potential in certain regions. The extension of Green Revolution technology to rainfed and dry regions, the neglect of small surfacewater harvesting systems such as tanks, and decline in public investment in irrigation in the 1990s have together contributed to the growing reliance on groundwater resources.

Dependence on groundwater has emerged as the single largest source of irrigation, with all its accompanying problems of serious risks to farmers' investment and degradation of environment.

Groundwater exploitation, however, has been uneven. There has been overexploitation in the dry regions, leading to serious and unsustainable depletion in these regions, and there are regions with high potential where groundwater remains underutilized due to the availability of cheap canal water. The existing irrigated areas have been displaying serious water stress as both reservoir and groundwater resources are depleting in many parts of the country. Cheap or free power alone may not be the cause for crisis, because farmers who invest huge sums on wells and borewells may not be deterred by power charges that form only a small fraction of capital costs like the high interest payment for borrowings from informal sources. Even with

Table 1.6

Net Area Under Irrigation by Sources in India

(in '000 hectares)

Year	Canals	Tanks	Wells and Tube Wells	Others	Total	NIA/ NSA (%)
1950–1	8,300	3,600	6,000	3,000	20,900	17.56
	(39.71)	(17.22)	(28.71)	(14.35)	(100)	
1970–1	12,838	4,112	11,887	2,266	31,103	22.17
	(41.28)	(13.22)	(38.22)	(7.29)	(100)	
1980–1	15,292	3,182	17,695	2,551	38.720	27.66
	(39.49)	(8.22)	(45.70)	(6.59)	(100)	
1990–1	17,453	2,944	24,694	2,932	48,023	33.41
	(36.34)	(6.13)	(51.42)	(6.11)	(100)	
2003–4	15,145	1,943	35,265	2,752	55,105	40.06
	(27.48)	(3.53)	(64.00)	(4.99)	(100)	
Growth Rate (per cent per annum)						
1970s	1.76	–2.53	4.25	1.19	2.21	
1980s	1.33	–0.77	3.00	1.40	2.18	
1990–1 to 2003–4	–1.09	–3.15	2.80	–0.49	1.06	

Note: Canals include both government and private, but the latter's share is negligible. NIA denotes Net Irrigated Area, NSA denotes Net Sown Area. NIA/NSA for 2003–4 is for the year 2002–3. Growth rate is compound annual growth rate.

Source: Ministry of Agriculture, Government of India (as reported in www.indiastat. com).

some kind of pricing mechanism, water use efficiency for irrigation would remain an important issue.

Unlike irrigated agriculture, rainfed agriculture is characterized by low levels of productivity and low input use. Further, variation in rainfall results in wide variations in instability in yields. A bulk of the poor in India lives in rainfed regions. Hence, it is important to give high priority to sustainable development of these areas through watershed development. Though the importance of watershed development has been recognized for long, nevertheless it has not made much headway except in a few pockets. The programmes are caught in bureaucratic muddle. An added problem is that the traditional water harvesting structures have become virtually defunct.

Environmental Stress

A serious source of environmental footprint of agriculture is increasing pollution of river and canal water. Many of the rivers and lakes are getting contaminated from industrial effluents and agricultural run-off, with toxic chemicals and heavy metals, which are hard to remove from drinking water with standard purification facilities. Irrigation undertaken by polluted water can also seriously contaminate crops such as vegetables and fruits with toxic elements.

Soil erosion is the most serious cause of land degradation in India. Estimates show that around 130 million hectares of land (45 per cent of total geographical area) is affected by serious soil erosion through ravine and gully, cultivation of wastelands, waterlogging, shifting cultivation, etc. It is also estimated that India loses about 5310 million tonnes of soil annually. According to estimates of the National Remote Sensing Agency (NRSA), the degraded land increased in the 1980s by 7 million hectares from 11.31 per cent to 18 per cent of cultivable area (Chand 2006). The accumulation of salts and alkalinity affects the productivity of agricultural lands in arid and semi-arid regions that are under irrigation. The magnitude of waterlogging in irrigated command is estimated at 2.46 million hectares (Pingali 2005). Besides, 3.4 million hectares suffer from surfacewater stagnation. Injudicious use of canal water causes waterlogging and a rise in the water-table, which, if left uncorrected, eventually leads to salinization. Although irrigation and drainage should go hand in hand, the drainage aspect has not been given due attention in major and medium irrigation projects in the country. There has been waterlogging associated with many of the large reservoirs since their inception (Government of India, Ninth Five Year Plan).

Fertilizers and pesticides are important inputs for increasing agricultural production. Their use has increased significantly from the mid-1960s due to the Green Revolution technology. Over and unbalanced use of these chemicals is fraught with danger. Serious problems have arisen in the Indo-Gangetic Plain (IGP) because of the distorted ratio of application of nitrogen, phosphorus, and potassium (NPK). There has been excessive use of nitrogen with adverse effects on soil fertility (Venugopal 2004). This is partly the result of price differentials, and partly due to lack of knowledge among farmers about the need for balanced fertilizer use. The consequence is soil nutrient depletion that is a major cause of the stagnation of rice yields. This is especially true in areas that make concentrated use of fertilizers and pesticides.

Another serious problem emerging from heavy applications of nitrogen in rice and wheat is contamination of groundwater. Nitrate cannot be removed once it has entered the underground water system. Excessive phosphorous contaminates surface water. Other problems relate to deficiency of trace elements because of intensive cultivation. All these factors have combined together to reduce soil fertility. Proper crop rotation, judicious combination of organic and chemical fertilizers, and suitable agronomic practices will be helpful in this regard.

DECELERATION OF GROWTH IN AGRICULTURE

The growth of agriculture in terms of both gross product and output has visibly decelerated during the post-reform period compared to the 1980s. For example, the growth rate of agricultural GDP decelerated from 3.08 per cent during 1980–1 to 1990–1 to 2.57 per cent during 1992–3 to 2005–6 (Table 1.7). The growth rate for all crops taken together decelerated to 1.96 per cent during 1990–1 to 2000–1, compared with a growth rate of 3.19 per cent during 1980–1 to 1990–1.

The cause for concern is not merely the decline in the rate of growth of agricultural production, but also the decline in the growth rate of foodgrains, which fell from 2.85 per cent in the 1980s to 1.16 per cent in the 1990s, lower than the rate of growth of population of 1.9 per cent during the latter period. The 1990s were thus the first

Table 1.7

Growth of Sectoral Income and Per Capita Income

(1999–2000 prices)

Year	Agriculture	Industry	Services	GDP at Factor Cost	Per Capita NNP at Factor Cost
1980–1 to 1990–1	3.08	5.79	6.54	5.15	2.82
1992–3 to 2002–3	2.61	5.82	7.65	5.85	3.89
1992–3 to 2005–6	2.57	6.05	7.72	6.00	4.10
1950–1 to 2005–6	2.54	5.19	5.40	4.26	1.94

Note: Growth is Compound Annual Growth Rate, GDP dentotes Gross Domestic Product, NNP denotes Net National Product.

Source: CSO, *National Accounts Statistics*, various years.

decade since the 1970s in which the rate of growth of food production fell below the population growth rate. This was essentially due to the gradual decline in the growth of yield levels, especially of some food crops. While the annual yield growth rate for all crops taken together declined from 2.56 per cent during the 1980s to 1.09 per cent during the later period, for rice it decelerated from 3.47 per cent to 0.92 per cent, and for wheat from 3.10 per cent to 2.21 per cent. In the case of cotton, the yield growth rate went down from 4.10 per cent during the 1980s to –0.94 per cent during the 1990s, partly because of declining effectiveness of pesticides due to pest resistance (Bhalla 2007).

The consequences of reduced public investment in agriculture, specially in infrastructure such as irrigation, science and technology, rural roads, and market yards, are stagnation in irrigated area, deterioration in the quality of infrastructure, very slow progress in bio-technological research, and lack of technological breakthrough appropriate for rainfed and drought-prone regions, which form almost 60 per cent of the cropped area. All these factors are adversely affecting efficiency of crop production and leading to production crisis in agriculture. The most important manifestations of the crisis are deceleration of agricultural growth combined with increasing input use inefficiency, thereby adversely affecting the profitability of agricultural production. These add credibility to the argument that liberalization policy measures in agriculture and the negative effects following from Green Revolution technologies together have contributed to the productivity crisis in agriculture (Vakulabharanam & Motiram 2007). We shall return to further elaboration on these aspects in the following section.

Wide Regional Disparity in Productivity and Growth

Regional disparity in agricultural development can be measured in many ways, namely variations in the levels of output, agricultural income, growth rates of agriculture, and per worker productivity in agriculture. Here, attention is drawn to variations in state-wise differences in per worker GDP in agriculture since this measure gives a rough idea about the variations in levels of living of agricultural workers across states. There are very large variations in per worker productivity in agriculture across states. For instance, Punjab's worker productivity of Rs 35,000 during 2004–5 was 7.5 times that of Bihar. It is basically this difference in per worker productivity that explains the very large differences in the standard of living of agricultural workers across states. The growth rates of agriculture both in terms of GDP from agriculture and agricultural output (and yield) have also decelerated in most of the states. Except for the states of Bihar, Gujarat, and Orissa, there has been a deceleration in the growth rates of agriculture in all the other states during 1993–4 to 2003–4 compared with 1980–1 to 1993–4. The growth rates in these three states were very low and statistically insignificant.

The disparities become much sharper at the district level. There are marked inter-district variations in agricultural growth. Only about 20 per cent of about 500 districts in the country contribute substantially to growth. An equal percentage have had stagnant yields for decades (Bhalla & Singh 2001). Such diversity calls for agricultural strategies designed on the basis of specificities at a much lower level of disaggregation that even the present agro-climatic zones are too large to capture the diversity.

POLITICAL ECONOMY OF TECHNOLOGICAL TRANSFORMATION AND DIFFERENTIATED AGRARIAN STRUCTURE

By the 1990s small–marginal farmers, with 80 per cent of the holdings and about 40 per cent of the cultivated land, came to numerically dominate Indian agriculture but yet ended up in a precarious position because of technological as much as policy changes over which they had no control due to lack of adequate political power. It is increasingly clear that the introduction of Green Revolution technology in its first phase during the 1960s and 1970s excluded the small–marginal farmers not only because of its limited spread, but also because of the latter's inability to access the resources required. While land reforms failed to bring about any radical redistribution of land, the introduction of new technology has brought about sharper differentiation among peasantry by opening up more profitable opportunities in agriculture to the rich peasantry. The impression created is that the high-yielding variety (HYV) technology, unlike the heavy farm machinery-based technology, does not impose any size barriers of entry for small–marginal farmers. But in actual practice, the 'new technology' has been biased in favour of those who have better command over resources. Even studies which found that the new technology benefited all, emphatically observed that 'the gains of larger farmers were disproportionately large' (Rao 1975). The rich peasants, with better resources, access to cheap credit and right information, better risk-bearing and input-buying capacity, and better reach to scarce inputs through cooperatives or the government agencies, have become adept in adopting the new technology (Byres 1994). Some of the middle peasantry who could access all the resources seem to have become rich themselves by adopting the new technology. But the small peasantries are caught in this milieu of struggling to adopt to the change, and to an extent succeeded in adopting HYVs, mainly in agriculturally well-developed regions that follow the rice–rice or rice–wheat cropping pattern (Venugopal 2004). Notwithstanding constraints of small farmers, rapid diffusion of yield improving technology was witnessed in the 1980s (Sen & Bhatia 2004) mainly in food crops like HYVs of rice and wheat.

Even as the spread of the Green Revolution through HYVs of food crops showed clear signs of adverse effects of soil degradation, water pollution, and deceleration in growth rate of productivity, there still remained vast rainfed and drought-prone areas inhabited by relatively resource poor farmers who required improved appropriate cropping strategies. Instead, the Green Revolution technology evolved to suit irrigated agro-ecological systems was extended to rainfed and drought-prone agro-ecological regions as well (Rao 2004; Ray 2007). The spread of 'lagged' Green Revolution (Harris-White & Janakaranjan 2004; Vakulabharanam & Motiram 2007) to semi-arid areas, to non-food crops, and upwardly mobile medium, small, and marginal farmers coincided with a number of technological and institutional changes that brought the farming community into a vortex of growing vulnerability.

First, farming systems have been undergoing a shift from crops based on traditional variety to HYVs to hybrids to genetically-modified (GM) crops. The shift to high value but high-risk hybrid and GM crops is also accompanied by increasing exposure to market dependence for seeds, which may also carry sui generis kind of intellectual property right for which farmers end up paying an exorbitant price. These changing cropping systems necessitate knowledge-based practices as well as timely and comprehensive extension systems and services. There have been significant changes in the structure of costs of production for more than two decades, reflecting the changes in technology and relative prices of inputs (Sen & Bhatia 2004). The cropping pattern has also changed from the cereal bias, particularly from coarse grain, to more diversified and high value crops, because of demand-driven factors such as fast changing composition of the food consumption basket. The answers, in the form of a shift towards organic farming and crop rotation, are not easy options, particularly to small farmers, unless accompanied by appropriate institutional support systems.

Second, there has been growing pressure on the resource base. More challenging is the pressure on water resources. While waterlogging, salinity, and reduction of wastage are some of the problems often addressed in the context of command areas, the serious crisis exists as a consequence of overexploitation and the resulting irreversible depletion of groundwater resources, especially in dry and drought-prone areas. The neglect of public investment in irrigation is one of the important reasons for the growing dependence on groundwater as a source of irrigation. It involves farmer-based private investment, which is obtained at very high interest rates from non-institutional sources by small farmers. The risks are high and the failure or depletion of the water table has often been the trigger for many suicides in these dryland regions. A study by activist scholars in Andhra Pradesh draws attention to the fact that in 2005, 57 per cent of the total irrigated area in the country was dependent on groundwater resources and it is likely to increase to unsustainable levels of over 70 per cent, which would spell high risk and ruin to farmers as much as power utilities (Narendranath et al. 2005), if the state does not intervene with increase in investment in irrigation.

Third, as a result of neoliberal reforms, the institutional sources of credit, and other inputs like fertilizers, pesticides, and seeds have been on the decline over the years. Most of the small–marginal farmers are driven to depend on unregulated sources. While the emerging system of agriculture necessitates more sophisticated knowledge-based information systems, extension systems are almost defunct or in disarray.

Fourth, while it is well recognized that a reasonable livelihood for a small–marginal farm could be ensured if part or even half of the income sources are in non-farm activities, there is very little evidence of this fact forming the basis of rural development strategy. There is considerable experience of other countries in this regard, but without much impact on the strategy in India. The shift from promotional rural industrial strategies to emphasis on the State playing a facilitator's role has further worsened the prospects.

Fifth, the challenge of the shift towards free trade–driven agriculture, even under the Agreement on Agriculture (AOA) of the World Trade Organization (WTO), does not demand indiscriminate abandoning of State support systems. There has been focused strategy of cost reduction based on State supported research and extension investments even in North America and Europe. A small-farmer based agriculture, if it should not only face the challenges of the competition but also rise to the possibility of export orientation, needs strong research and extension support in the whole range of areas from developing appropriate plant varieties to sustainable resource use to market facilitation. But on all these counts the post-reform experience has been one of retrogression in state initiatives interspersed with outright privatization of certain activities such as development and distribution of seeds.

By the late 1980s, having exhausted the initial productivity gains, all that Green Revolution technology left in better resource endowed areas was plateauing of productivity with much degradation of soil and water pollution, and in resource poor areas spread of inappropriate technology with unprecedented risks in farming. The eagre late entry of small–marginal farmers in better resource endowed areas and that of entire farming communities in rainfed or dry regions into the mould of the so-called 'lagged Green Revolution' was already a potential disaster. At this juncture, the initiation of economic reforms and the decline of State support systems unleashed serious agrarian crisis. Ironically, a thin upper crest of rich farmers who gained through the initial benefits of the Green Revolution, with hefty State support and subsidies, appear to have had hardly any commitment to agriculture by late 1980s. Based on a longitudinal study of villages in Tamil Nadu, Harris-White & Janakaranjan (2004:414) provide a graphic description of how the upper crest of capitalist class emerging from agriculture has diversified its interests, which is by and large the same for much of the country:

Diversity is the hallmark of the expansion of rural capitalism here. It is agrarian households with the larger land holdings and hired labour forces which not only diversify into both income-elastic and water sparing agricultural products but also (because of ceiling on the absorptive capacity of agriculture and because of higher rates of return) into non-farm economy. Such assets accumulation, both within the village economy and directly and indirectly outside it, renders these households doubly diversified. When combined with salaried employment in the state or the urban economy (as happens in these households), they are trebly diversified. Such diversification is a close associate of agrarian differentiation...this class is endowed with risk-resisting economic plasticity.

This class has very little economic stake in agriculture but holds on to agriculture for nurturing their political ambitions that ensure much more economic gain through contracts and commission agencies than agriculture per se. With this political class representing agriculture, the reform process unfolds much easily than one would expect, as we shall see below.

REFORMS AND IMPACT ON THE FARMING COMMUNITY IN AGRICULTURE

The roots of the present all-pervading crisis in Indian agriculture could be traced back to the complacency and gross neglect of agriculture since the mid-1980s. Agriculture had fallen from policy priority under the euphoria that the country had left behind the days of shortages and achieved sustainable self-sufficiency in foodgrain production; that agriculture had attained a level of development where it could respond to the domestic market as well as global prices, if only the market restrictions are reversed; and that preferential and institutional interventions are anachronisms. But a worse deal had to wait till the 1990s since the unfolding economic reforms overwhelmed the peasant stability. The reforms influenced every measure of public policy, including agriculture, at the behest of the Union Government, and were carried on with different degrees of zeal at the state level. The unfinished distributive land reforms were seen as obstacles to incentives, and liberal markets were expected to bring about technological breakthrough. The result is rapid decline of institutional support to agriculture based on well-deliberated principles of growth with equity. The evidence compiled here, at both the macro and micro level, suggests the rapid retrogression in the public agricultural support systems manifesting in unprecedented stress on the farming community, resulting in widespread suicides.

Crisis in agriculture, which was well underway by late 1980s because of loss of priority in public provision, was deepened by the economic reforms beginning with the 1990s and has become all pervasive. The manifestation of the crisis is felt in different forms in different agro-climatic and institutional contexts. For instance, the spread of inappropriate Green Revolution technology as the only alternative even to rainfed and dryland ecological systems had severe adverse effects. As observed earlier, in the absence of surface water irrigation facilities, farmers in dry regions had no choice but to incur serious debts by investing in unstable groundwater resources. On the other extreme, the growing pressure on land in irrigated command areas resulted in a rapid increase in the highly exploitative tenancy system. The volatile prices of commercial crops, including certain plantation crops, have suffered ruination because of the agricultural trade liberalization. The exposure to externally engineered crops with a hope of very high yields, with very scant regard for their suitability to domestic conditions have resulted in inappropriate technological practices that meant severe loss of not only livelihoods but also resource degradation.

Trade Liberalization

In 1991, when India officially went along the structural adjustment path and introduced a series of neoliberal economic reforms, there was apparently not much explicitly by way of reforms in agriculture. But very soon, at least by the mid-1990s when the WTO was in place, there did unfold many policy reforms directly addressed to agriculture. The entire agenda of trade reforms in agriculture was carried on under the doctrine that border prices for agricultural commodities are higher than Indian prices and the farmers benefiting from these prices would invest their profits in agriculture, contributing to higher productivity and growth. In other words, getting the prices right would do the trick for agricultural development. Table 1.8 lists some of the important policy changes and measures of reform relating to Indian agriculture. Beginning with 1997, international trade in agriculture is liberalized. All Indian product lines are placed under Generalized System of Preferences (GSP). By 2000 all agricultural products are removed from Quantitative Restrictions (QRs) and brought under tariff system. Canalization of trade in agricultural commodities through state trading agencies was almost removed, and most of the products were brought under Open General Licensing (OGL). The average tariffs on agricultural products, which stood at over 100 per cent in 1990, were brought down to 30 per cent by 1997 and were targeted to come down further.

Pressure on Subsidies and Support Systems

Internally, the structural adjustment process initiated in 1991 at the behest of the International Monetary Fund (IMF), and pursued with the aid of the World Bank, had far-reaching implications for Indian agriculture. The single-minded pursuit of fiscal reforms had much greater effect on the agricultural input support system and institutions than even the provisions of the AOA of the WTO. Much of the Green Revolution initiated in the 1960s in India was built with a system of state supported incentives or subsidies and public investment in agricultural infrastructure like irrigation. The National Seed Corporation established in 1963, and later, a network of State Seed Corporations established since 1975 had virtual monopoly and responsibility of developing and distributing better and HYV seeds in collaboration with the agricultural universities. Though opening up of trade in seeds and the seed industry to private trade and foreign investment started in a limited way under the Seed Development Policy 1988 (Venugopal 2004) by 1991, 100 per cent foreign equity was allowed in the seed industry in India and restrictions on import of seeds were relaxed.

Table 1.8

Important Measures of Economic Liberalization in Indian Agriculture

Area of Liberalization	Policy Changes and Measures of Implementation
I. External trade sector	a. In tune with the WTO regime, since 1997 all Indian product lines placed in Generalised System of Preferences (GSP).
	b. India is a part of the WTO intellectual property rights (IPR) regime relating to agricultural products including plant varieties (seeds) and Geographical Indications.
	c. In 1998, quantitative restrictions (QRs) for 470 agricultural products dismantled. In 1999, further 1400 agricultural products brought under Open General Licensing (OGL) and canalization of external trade in agriculture almost reversed.
	d. Average tariffs on agricultural imports reduced from 100 per cent in 1990 to 30 per cent in 1997.
	e. Though India in principle is against Minimum Common Access, but actually already importing 2 per cent of its food requirements.
	f. More liberalized imports of seeds.
II. Internal market liberalization	
1. Seeds	a. Since 1991, 100 per cent foreign equity allowed in seed industry.
2. Fertilizers	a. Gradual reduction of fertilizer subsidies since 1991.
3. Power	a. Since 1997, power sector reforms were introduced.
	b. Power changes to agriculture increased but there is resistance from state governments

like Andhra Pradesh, which have introduced 'free power' to agriculture.

4. Irrigation	a. Water rates increased in some states.
	b. Participatory water management was sought to be introduced through Water Users' Associations (WUAs).
	c. States like Andhra Pradesh made new large irrigation projects conditional on 'stakeholder' contribution to part of investment but later government ignored this condition and increased public investment in irrigation.
5. Institutional credit	a. Khusro Committee and Narasimham Committee (1992) undermining the importance of targeted priority sector lending by the commercial banks. Targets for agriculture were allowed to be compromised.
	b. A number of bank branches in rural areas were closed.
	c. The objectives of Regional Rural Banks' (RRBs) priority lending to weaker sections in rural areas diluted since 1997 and RRBs are being restructured on commercial considerations.
6. Agricultural marketing	a. Changes in the provisions of Essential Commodities Act.
	b. Relaxation of restrictions on the inter-state movement of farm produce.
	c. Model Agricultural Market Act.
	d. Encouragement of contract farming.
	e. Agricultural commodity forward markets.
III. Fiscal reforms	a. Fiscal reforms with an emphasis on tax reduction and public expenditure tuned to reducing fiscal deficit as priority. (Grave implication for public investment in agriculture and rural infrastructure.)

Sources: Acharya (2004: 677), Chand (2006), Dorin and Jullian (2004: 206), and Vakulabharanam (2005: 975).

Fertilizer subsidy, which continues the major explicit agricultural incentive system directly funded by the Union Government of India, has been considerably reduced. Fertilizer subsidy, which amounted to 3.2 per cent of GDP and 6 per cent of the Union revenue expenditure in 1990–1, was reduced to 2.5 per cent and 5 per cent respectively by 1997–8 (Acharya 2004: 67). It was further reduced to 0.69 per cent of GDP by 2003–4 (Sen & Bhatia 2004: 275). Low electricity charges for agriculture are an incentive system provided through state budgets. Since 1997 several state governments introduced power sector reforms at the behest of the World Bank, and increased power tariffs with the ultimate objective of cost recovery. As part of the reforms, the power sector was thrown open to private sector investment. Low water rates for irrigation have been yet another implicit incentive to farmers provided through the state budgets. Many states revised the water rates upwards with the objective of recovering operation and maintenance costs. Some states, for example Andhra Pradesh, had announced a ban on investment on new major irrigation projects, unless the 'stakeholders' also contributed to part of the investment. But it is a different matter that the later government in 2004 ignored these conditions and went ahead with a programme of huge investment in irrigation.

Growing Informal Credit

Even at the risk of repetition, it must be emphasized that substantial proportion of Indian agriculture is a 'small farm' based economic activity, and is increasingly moving from a system of farmers' own-resource-based subsistence farming to purchased-input-based intensive commercial farming. Further, since small farmers' own resources are much too meagre, timely and assured credit at reasonable interest rate has become a critical input in Indian agriculture. In the face of inadequacy or non-functioning of agricultural cooperatives, part of the radical banking reforms of the earlier times in the 1960s in the form of 'social control', and later by way of bank nationalization, were aimed at increasing the flow of institutional credit to agriculture by prioritizing lending to this sector. But beginning with 1991, at the behest of pressures from the reform agenda, 'targeted priority lending' or 'directed credit' to agriculture was put on the back burner. The Narasimham Committee on the Financial Reforms (1992) recommended the dilution of priority sector lending, including lending to the agricultural sector by commercial banks. Though for political reasons, there was no explicit policy of removing priority lending to agriculture, the insistence on adherence to commercial performance placed a severe constraint on bank credit to agriculture. Instead of expanding rural bank branches, there was actually closure of rural bank branches, which declined from 34,867 in 1990 to 32,386 in 2003 (Rao 2004b). The Regional Rural Banks (RRBs), which were meant for lending specifically to 'weaker sections', were opened to all on commercial principles with an upward revision of interest rates (Rao 2004a).

The scheduled commercial banks' share of credit to agriculture declined from 18 per cent in December 1987 to 11 per cent by March 2004 (Shetty 2006). A study of credit from formal institutional sources shows that between 1980–1 and 1999–2000, the agricultural sector's share of short-term credit declined from 13.3 per cent to 6.1 per cent (Rao 2004b). The acceleration in the decline in the share of much needed long-term credit for investment was witnessed since the early 1990s. The number of agricultural loan accounts in scheduled commercial banks that had reached a peak of 27.7 million by March 1992 declined to 20.3 million by March 2002 and stood at 21.3 million by March 2004 (Shetty 2006). The worst sufferers of the formal institutional resource crunch have been the small borrowers, who are mostly small farmers. Beginning with the early 1990s, especially since 1993, the small borrowers' share in bank credit declined steeply from 21.9 per cent in 1992 to 7 per cent in 2001 (Rao 2004b). This does not mean that small farmers' needs have gone down or they were restrained from borrowing. It only means that small farmers were forced to borrow from non-institutional sources such as moneylenders, fertilizer and pesticide dealers, friends, and relatives. The interest charges of these informal sources are disproportionately high as compared to institutional credit.

A recent nationwide survey (NSS 59th Round, Reports 498 and 501, 2005) also brings out the grave agrarian situation in terms of farmer indebtedness. While almost 50 per cent of the farming households are indebted, the percentage is much higher in Andhra Pradesh (82.0 per cent), Tamil Nadu (74.5 per cent), Punjab (65.4 per cent), and Kerala (64.4 per cent), which are also states with relatively higher investment. More than 50 per cent of the borrowing is for investment in agriculture, the percentage being much higher in Andhra Pradesh (77 per cent), Karnataka (73 per cent), and Maharashtra (83 per cent). Institutional sources account for about 50 per cent on an average, with the share being much lower at 30 per cent in some states, as for example Andhra Pradesh, where the rest 70 per cent comes from informal sources. While interest charges on institutional credit are less than 20 per cent in 98 per cent of the cases, interest charges on informal credit are more than 20 per cent in more than 74 per cent of the cases.

Declining Public Investment

The economic reforms in Indian agriculture intensified the process of public as well as private resource crisis brewing from the mid-1980s. Gross Capital Formation (GCF) in Indian agriculture declined drastically. The public sector GCF in agriculture declined to one-third in 1999–2000, of what it was in 1980–1 (Reddy 2006b). Contrary to the expectations, the reform measures did not stimulate much increase in private investment. On the contrary there was deceleration of growth of private investment in agriculture (Sen 2003) and as a result, the overall GCF in agriculture as a share of total capital formation in the country declined almost by half, during the 1990s, from 13.1 per cent to 7.4 per cent. The proportion of plan expenditure on agriculture and allied activities declined from 6.1 per cent to 4.5 per cent (Mishra 2006a). Table 1.9 shows that there was a steep decline in the share of public sector in GCF in agriculture from 43.2 per cent in 1980–1 to 22.6 per cent in 1990–1, and further down to 15.5 per cent in 2002–3. There was a marginal increase in the recent years to about 19 per cent. Further, there was a drastic reduction in the share of developmental expenditure on rural development from 11.7 per cent of the GDP in 1991–2 to 5.9 per cent in 2000–1 (Gupta 2005: 5).

Rising Costs

There has been steep increase in the costs of farming across the country, which is substantially due to reforms. The fertilizer price index increased from 99 in 1990–1 to 228 in 1998–9 at a compound annual

Table 1.9

Capital Formation in Agriculture

(1999–2000 series)

Year	Agriculture and allied activities (Rs crore)			GCF in agriculture as per cent of agricultural gross domestic product (GDP)		GCF in agriculture as per cent of aggregate GDP	GCF in agriculture as a per cent of aggregate GCF
	Total gross capital formation (GCF)	Public sector GCF	Private sector GCF	Public	Private		
1980–1	4,342	1,876 (43.2)	2,466 (56.8)	4.0	5.2	3.0	16.1
1985–6	6,364	2,807 (44.1)	3,557 (55.9)	3.5	4.5	2.3	9.6
1990–1	15,839	3,586 (22.6)	12,253 (77.4)	2.4	8.1	2.8	11.5
1993–4	16,230	4,874 (30.0)	11,356 (70.0)	2.1	5.0	1.9	8.8
1994–5	17,392	5,952 (34.2)	11,440 (65.8)	2.3	4.3	1.7	7.3
1995–6	19,838	6,678 (33.7)	13,160 (66.3)	2.3	4.6	1.7	6.3
1996–7	24,107	7,214 (29.9)	16,893 (70.1)	2.1	4.9	1.7	7.9
1997–8	28,701	6,779 (23.6)	21,922 (76.4)	1.9	6.0	1.9	7.9
1998–9	31,021	7,476 (24.1)	23,545 (75.9)	1.8	5.6	1.8	7.8
1999–2000	50,151	8,670 (17.3)	41,481 (82.7)	1.9	9.3	2.6	9.8
2000–1	46,432	8,176 (17.6)	38,256 (82.4)	1.8	8.5	2.2	9.2
2001–2	60,366	10,353 (17.2)	50,013 (82.8)	2.1	10.3	2.6	11.1
2002–3	61,883	9,564 (15.5)	52,319 (84.5)	2.0	11.1	2.5	10.1
2003–4	61,827	12,218 (19.8)	49,609 (80.2)	2.3	9.3	2.2	8.4
2004–5	70,786	13,610 (19.2)	57,176 (80.8)	2.5	10.7	2.3	7.6
2005–6	83,952					2.4	7.3

Note: GDP at current market prices. Figures in parentheses indicate percentages to total GCF in agriculture.

Source: CSO, *National Accounts Statistics*, various years.

growth rate of 11 per cent (Acharya 2004: 73). And one estimate, across the crops and country, suggests that fertilizers presently account for 29 per cent of farmers' input costs (Acharya 2004: 78). There have been increases in the water charges in many states. One of the often cited reasons for agricultural trade liberalization is that it provides access to higher prices in the global markets. However, there has actually been a decline in global prices of some of the agricultural commodities such as rice and cotton for which India enjoyed comparative advantage. Before 1998–9, the Indian domestic lint prices were lower than world prices and India was an exporter of cotton. With the removal of QRs and with the fall in the global cotton prices in 2001–5, India has turned into an importer of cotton, which depressed domestic prices of cotton and has been the cause of serious losses to cotton farmers (Vakulabharanam 2005). According to one estimate, most of the global agricultural commodity prices in 2002 were lower than those in 1994, and in particular, cotton prices were 30 per cent less (Vakulabharanam 2005). The trade scenario in agricultural commodities after 1991 reflects the impact of economic liberalization and steep devaluation of the Indian rupee in the 1990s. Although the country was able to accelerate the growth rate of agricultural exports, the boon was short-lived. After 1997, there was deceleration in the growth of agricultural exports. International prices of most of the commodities started falling and made Indian exports non-competitive (Government of India 2007). Contrary to gains from trade in agriculture, there were severe adverse effects on domestic agriculture because of the liberal imports, especially of edible oil and spices.

Adverse Terms of Trade and Declining Farm Business Income

Another important manifestation of crisis in agriculture is the stagnant, if not deteriorating, terms of trade for agriculture after the introduction of economic reforms. Barter terms of trade became favourable to agriculture up to 1996–7 and almost stagnant thereafter (Government of India 2007). The Planning Commission, in its *Mid-Term Appraisal of the Tenth Five Year Plan*, had to admit to the grim state of agriculture, when it observed: 'During 1997–2002, agricultural prices declined relative to prices not only of inputs but also non-food consumer goods. As a result, purchasing power of agricultural incomes (current price GDP deflated by consumer expenditure deflator) decelerated more than GDP at constant prices. Real farm incomes defined in this way not only show no per capita growth after 1996–7, but also increased variability' (Government of India 2005).

Farm Business Income (FBI), the difference between the value of output produced and the costs actually paid out, which was on the rise in the 1980s, started declining in the 1990s. The growth of FBI per hectare decelerated from 3.21 per cent in the 1980s to 1.02 per cent in the 1990s. 'The growth of real FBI per cultivator declined from 1.78 per cent in 1980s to 0.03 in 1990s and in actual terms also it seems to have declined in the states of Andhra Pradesh, Bihar, Gujarat, Karnataka, Maharashtra, Orissa and Rajasthan' (Sen & Bhatia 2004: 42).

Figure 1.1 shows the steep rise in the cost of living in rural areas as indicated by the Consumer Price Index for Agricultural Labour (CPIAL) while the farmers' income languishes. This is a familiar

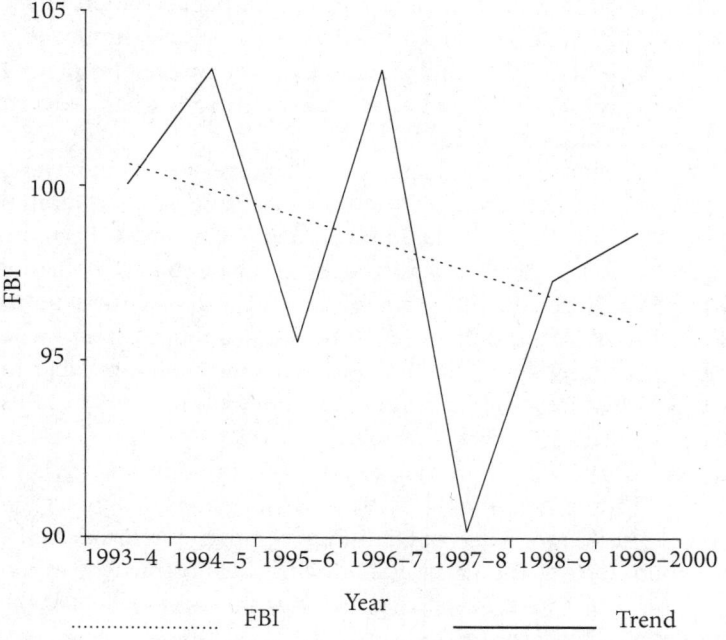

Figure 1.1: Indices of Farm Business Income (FBI) deflated by Consumer Price Index for Agricultural Labour (CPIAL)

Table 1.10
Per Worker Income in Agriculture and
Non-agriculture Sectors in India

(1993–4 prices)

Period	Income per worker (Rs)		Ratio of non-agriculture to agriculture	Decadal growth rates (%)	
	Agriculture	Non-agriculture		Agriculture	Non-agriculture
1978–9 to 1983–4	9,961	28,430	2.85	–	–
1988–9 to 1993–4	11,179	39,355	3.52	1.16	3.31
1998–9 to 2003–4	11,496	59,961	5.22	0.28	4.30

Source: Chand (2006).

scissors crisis in agriculture, often resulting in pauperization of the peasantry. This has also resulted in widening of disparities between agricultural and non-agricultural incomes as can be seen from Table 1.10. The disparities have doubled over the last two and a half decades, leaving agriculture way behind.

If consumption expenditure based headcount estimates of poverty are used, one may not perceive the stress on agricultural communities but by looking at the undernourishment, the stress becomes apparent. Table 1.11 gives data separately on the number of poor and undernourished persons in various farm categories in rural India. What is of significance is that even as the headcount of persons who are poor is coming down, there has been a spurt in the number of undernourished across all farming classes, especially in the 1990s. This is also true when 2004–5 is compared with 1993–4, as consumption estimates for 1999–2000 are not comparable. In addition, one also observed a shift of households from marginal farmers to agricultural labourers. This clearly brings out the adverse impact of reforms on the health conditions of the farming community. Unfortunately, we do not have data that would capture undernourishment in terms of the health status of the farming community.

Table 1.11

Number of Poor and Undernourished Persons in Various Farm Categories in Rural India

(in million)

Year	Agricultural farm classes		Marginal (<1 ha)		Small (1–2 ha)		Semi-medium (2–4 ha)		Medium (4–10 ha)		Large (>10 ha)	
	Poor	Under-nourished	Poor	Under-nourished	Poor	Under-nourished	Poor	Under-nourished	Poor	Under-nourished	Poor	Under-nourished
1983–4	44.6	33.7	131.2	98.0	41.1	25.8	29.5	18.0	15.0	9.2	2.8	1.9
1987	40.0	30.2	115.1	84.0	29.6	18.8	16.6	12.3	7.2	5.3	1.2	0.7
1993–4	39.5	39.2	123.5	105.5	26.7	24.7	15.0	12.4	8.4	7.4	0.8	1.0
1999–2000	36.5	42.8	95.2	122.0	16.4	28.7	8.5	18.7	3.2	10.3	0.0	0.7
2004-5	80.4	156.5	31.6	82.2	17.9	57.8	9.3	40.7	3.5	17.1	0.3	2.5

Sources: Kumar (2005) for 1983–4 to 1999–2000, and own estimates based on unit-level data for 2004-5.

AGRARIAN CRISIS AND FARMERS' SUICIDES

In addition to rising costs, volatile prices, growing risks, vanishing support systems, etc., that have resulted in the present crisis in agriculture, there has been increasing pressure on the farmers in terms of meeting basic social needs such as education of children and health of family members, which are increasingly privatized and are becoming a considerable part of domestic expenditure needs. A combination of these stress factors has been at the root of the unusual phenomenon of farmers' suicides in rural India, especially since 1997. Though there are limitations of data on suicides, an attempt is made here to look into the nature of sources, the nature of data and, to the extent possible, use the same in understanding the links between post-reform agrarian crisis and the worst form of its manifestation in the form of suicides.

The main official source of data on suicide deaths is police records made available by the National Crime Records Bureau (NCRB), Ministry of Home Affairs, Government of India. The limitation of the annual data provided by the Bureau is the routine reporting of suicides, which may not reflect the current crisis in agriculture. Further, suicides are likely to be underreported because the act is identified with shame and stigma and also because of a legal sanction against it.[1] Notwithstanding these limitations, attempts have been made to analyse the trends in mortality, suicide mortality, and farmers' suicides at the district, state, and national levels (Mishra 2006a, 2006b, 2006c, 2006d; Mohan Rao 2004).

While the NCRB data are available from 1975, the profession-wise distribution is published only from 1995. The available data show that even as the overall Crude Death Rate (CDR) has been coming down, suicide mortality rate (SMR) has been on the rise for the country as a whole (Mishra 2006c). Since most of the suicides are among male farmers, it would be interesting to focus on male SMR in India as well as in states reporting high incidence of farmers' suicides. The male SMR is much higher than the overall SMR. The rate of growth of male SMR was much higher in the 1990s than earlier. The male SMRs of Andhra Pradesh and Maharashtra which were close to the national average till the late 1980s, started rising at much faster rate in the 1990s. These two states are among the four states that have reported the highest

Table 1.12
Age-adjusted SMR for All Males and for Male Farmers

Year	India		Andhra Pradesh		Karnataka		Kerala		Maharashtra		Punjab	
	All Males	Farmers Male	All Males	Farmers Male	All Males	Farmers Male	All Males	Farmers Male	All Males	Farmers Male	All Males	Farmers Male
1995	12.5	10.5	11.4	13.6	31.8	33.6	42.0	127.4	17.4	14.7	4.4	5.2
1996	11.9	12.2	13.3	24.4	24.8	30.9	40.2	109.3	16.0	23.5	4.1	7.3
1997	12.9	12.7	14.8	17.5	28.1	31.3	45.7	138.8	17.7	23.9	4.0	6.1
1998	13.8	14.8	16.6	28.8	30.0	30.1	47.4	172.7	18.9	29.0	5.3	5.9
1999	14.4	15.3	18.1	30.0	33.4	41.4	49.6	182.4	18.5	30.6	6.6	4.8
2000	14.2	15.7	17.4	22.8	33.2	43.5	47.4	184.6	19.6	37.3	5.5	4.1
2001	14.0	16.2	18.2	25.6	32.3	44.5	48.3	161.8	20.6	44.1	3.4	2.4
2002	14.3	18.1	21.2	31.8	32.6	41.6	50.5	258.5	20.3	47.3	3.3	2.3
2003	14.5	18.0	20.7	28.5	33.2	48.3	48.5	298.0	20.6	50.8	4.1	1.5
2004	14.4	19.2	24.7	44.6	31.2	35.5	45.8	183.4	20.3	57.3	4.1	4.3
2005	14.1	18.3	23.7	41.2	29.5	34.7	47.0	249.3	19.3	55.1	3.8	2.8
2006	14.4	18.2	23.4	41.6	30.2	30.4	44.9	260.4	20.6	62.6	4.7	5.1

Note: SMR denotes Suicide Mortality Rate (suicide deaths per 100,000 per sons). Unavailability of data at all-India level leads to exclusion of (a) Andaman and Nicobar Islands, Pondicherry, Rajasthan, Sikkim, and Tamil Nadu in SMR for male farmers in 1995, (b) Pondicherry in SMR for male farmers in 1996, and (c) Jharkhand in both the categories in 2003.

Source: NCRB (various years), as in Mishra (2006c).

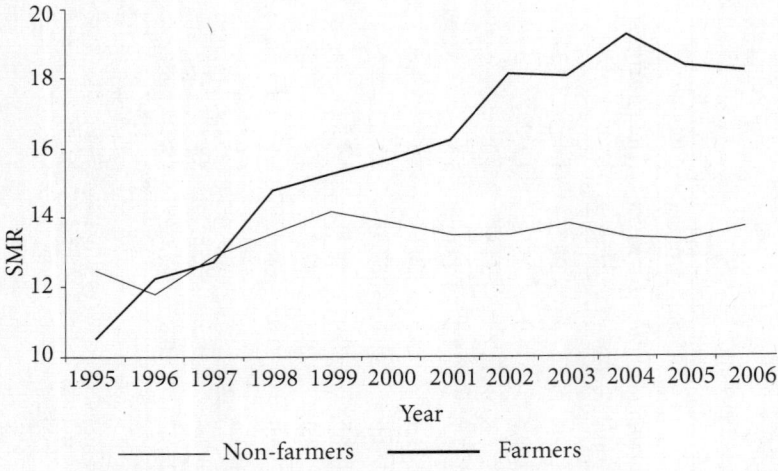

Figure 1.2: Suicide Mortality Rate for Male Farmers and Male Non-farmers in India: 1995–2006

Note: SMR denotes Suicide Mortality Rate. SMR for farmers/non-farmers is based on interpolated/extrapolated population for cultivators/non-cultivators using 1991 and 2001 census. For details, see Mishra (2006c). Due to data unavailability, SMR estimates exclude Andaman and Nicobar Islands, Pondicherry, Rajasthan, Sikkim and Tamil Nadu in 1995, Pondicherry in 1996, and Jharkhand in 2003.
Source: NCRB, various years.

incidence of farmers' suicides since the early 1990s. The other two states, Karnataka and Kerala, which had much higher SMRs than the national average up to the 1980s, show a further rise in the 1990s.

As observed earlier, profession-wise SMRs are available only from 1995. Figure 1.2 shows that SMR for male farmers has been rising steeply since 1995, while SMR for male non-farmers has been more or less stable, especially since 1999. Table 1.12 shows the age-adjusted SMRs for all male population as well as SMRs for male farmers for India and the five states under discussion. For India as well as the states of Andhra Pradesh, Kerala, and Maharashtra for the period, and Karnataka till 2003, the gap between male SMRs in general and male farmer SMRs has been on the rise. There has been a decline in this gap in Karnataka since 2004. In Kerala, the gap was very high and started widening further since 1997, though there was a decline in 2001 and again in 2004.

Regression and correlation results based on cross-section data of 1995 for 19 states in the country show that the rate of suicides of farmers are more manifest in areas with favourable ratio of area to holdings among small farmers, higher rate of suicide in the general public, high percentage of deaths to economic bankruptcy, and higher proportion of area under non-food crops. The results also show that the farmer suicides are in greater numbers in areas with predominance of small holdings, groundwater-based, and low share of priority sector advances to agriculture. In this connection, the negative association between area under cotton and share of rural credit in total institutional credit also deserves attention (Mohan Rao 2004).

Besides the NCRB, the second, and more widely used, source of data is press reports, particularly from 1997, which are based on the suicides specific to farming-related causes. The data based on press reports may have an element of overemphasizing the failure of institutional facilities because of linking it to all suicide deaths by farmers. Nonetheless, studies such as the one by Mohan Rao (2004) do provide some insights. The study made a content analysis of 337 media reports and brought out economic causes as the dominant factor for the suicides of farmers in Andhra Pradesh. Among the economic factors, indebtedness, crop failure, and lower prices are prominent though higher rates of interest and liberalization also figured. Among non-economic causes, harassment from creditors, particularly input dealers and moneylenders, emerged as a major factor, though spurious inputs, overuse of pesticides, and erratic power supply also figured. Next in order are institutional factors, namely lack of institutional credit, and limited purchases by Andhra Pradesh State Cooperative Marketing Federation Limited (MARKFED) and Cotton Corporation of India (CCI). Among the natural factors, inadequate rainfall during the sowing season and heavy rainfall at the time of harvest figured prominently.

The third source is the official data released by the state governments, but this suffers from underestimation because it is strictly linked to compensation paid by the government departments. Official scrutiny often is known to treat even genuine instances of suicide as arising out of disease or old age or some other reason, with a view to restrict the payment of compensation. On 10 April 2006, the Union Ministry of Agriculture convened a meeting of chief ministers and agriculture ministers from the four states that have been reporting high incidence of suicides. Statistics relating to farmers' suicides during the five years 2001–2 to 2005–6 were released. Officially, the total number of suicides during the period was reported as 5910 in Karnataka (Karnataka Government disputed it as an overestimate), 1835 in Andhra Pradesh (Andhra Pradesh Government disputed it as underestimate and corrected it as 2035), 981 in Maharashtra, and 201 in Kerala (*The Hindu*, 11 April 2006). Official confirmation of 9127 suicides by farmers in the four states in just five years is no mean statement, and it is the highest ever in post-independence India. However, this is much lower than the NCRB data, which indicates that between 2001 and 2006 there were 56,946 suicide deaths by farmers in these four states and 1,03,982 for the country as a whole.

Household Surveys

Notwithstanding these limitations, the available data on suicides do indicate their links to the impact of reforms on resource poor and small-farmer Indian agriculture, and the resulting rural distress. Although it is not an exemplar methodology for social science researchers to visit the households of the suicide victims with an investigative zeal, particularly when the households realize that these investigators are not even empowered to recommend compensation, still many small sample-based studies of the households of suicide victims do exist, to which we shall return in the next section. Though suicides are reported from many parts of the country, the magnitude varies.

This volume includes four studies (Chapters 6–9), each dealing with one of the four high farmer suicide incidence states, and based on household level suicides as well as secondary sources. These and other household level studies from across the states show that the farming community is passing through high stress during this high growth reform period. Though the specific and immediate triggers may vary from region to region, the overall macroeconomic context is one of structural adjustment and trade liberalization. The evidence available from the states does emphasize the reform-induced stress. In Andhra Pradesh, declining public investment in irrigation and unavailability of credit from institutional sources meant heavy investment in digging of borewells by borrowing from informal sources at a higher interest burden (AWARE 1998; CES 1998; Rythu Sahaya Committee [RSC]

1998; Shiva et al. 2000). The growing dependence on groundwater and competitive exploitation of the limited resources resulted in the depletion of the water table and failure of borewells, resulting in huge unredeemable debts. An enterprising and hard-working farmer is now identified as someone who is reneging on contracts—he cannot repay his loans. It is the government sponsored mission on oilseeds that led to the extension of Green Revolution technologies, with cropping pattern shifts towards oilseed mono-cropping, expecting higher returns (Vidyasagar & Chandra 2004). This not only reduced the farm-based risk mitigation available from mixed and multiple cropping cultivation in dry regions, but the post-reform developments in liberalizing imports of edible oil has also led to the crashing of domestic prices and returns even during normal times. It is observed that 55 per cent of farmers in Andhra Pradesh do not get minimum support prices (MSP) (Vidyasagar & Chandra 2004). The worst affected are the marginal, small, and even medium farmers who do not get the MSP because of their dependence on traders for credit not only at high rates of interest, but also tied up with the purchase of output at prices lower than the MSP. A study by Bhushan and Reddy (2004) based on a survey of suicide households indicates that the households had taken to mono-cropping of input-intensive non-food commercial crops and even leased in land, but crops failed due to inadequate water. Another study (Rao & Suri 2006) points out that increasing costs and low returns add to the crisis and this has come about because of neoliberal policies, which are in turn a consequence of the loss of power of the farming community, resulting in their removal from the policy-making process.

The rain-dependent cotton growing farmers of Vidarbha in Maharashtra are faced with declining profitability because of dumping in the global market by the USA, low import tariffs, failure of the Monopoly Cotton Procurement Scheme, and withdrawal of supporting state investment and subsidies (Mishra 2006a). Another study of suicides of farmers in the same region of Maharashtra in the Durkheimian framework also observes that 'lower and middle caste peasant smallholders found themselves trapped between enhanced aspirations generated by land reforms and other post-1947 measures, and the reality of neoliberalism' reflected in rising debt and declining income (Mohanty 2005).

The global exposure of the plantation-based farmer in Kerala also suffered because of depressed prices (Nair & Menon 2005). A study on farmers' suicides in Kerala concludes that farmers' distress over the past one decade is closely linked to the neoliberal policy regime in the 1990s. 'The association between the two is more in the regions of the state which are heavily dependent on export-oriented crops, such as coffee and pepper' (Mohanakumar & Sharma 2006). In drought-prone Karnataka, it is the liberal imports of edible oil, exposure to fluctuating agricultural commodity markets, and decline in public investment and state support systems to agriculture that triggered suicides (Assadi 1998; Deshpande & Prabhu 2005; Vasavi 1999; Vidyasagar & Chandra 2004).

REFORM-LED GROWTH, SMALL PEASANT ADJUSTMENT COST, AND NEED FOR STATE SUPPORT

As reports of farmers' suicides from parts of the country appeared in the mid-1990s, initially the response of the Central as well as the state governments was to characterize it as a kind of psychological aberration of the farming community. As the adverse effects of the reform process deepened, there is official recognition of the agrarian crisis (Government of India 2005a) and the need for intervention by the state to provide relief to the farmers (Government of India 2007, Ch. 4). True, the failure of the village as a social community and the growing disintegration of the joint family as a protective and supportive collective may have contributed to conditions that would make the triggers to operate easily. But the context is the agrarian crisis seen in terms of increasing input costs, decreasing profitability, volatile commodity prices, growing risks, and degradation and depletion of land and water resources. A critical question is whether small–marginal farming, especially in rainfed and dry regions, is sustainable without substantial public infrastructure support and comprehensive social security including health, education, employment, and old age support. Even at the early stages of the structural adjustment programme (SAP), there was clear warning that neoliberal reforms would face adjustment among poor farmers, which without assistance from the state, would intensify their suffering (Cornia et al. 1987). By and large, the incidence of suicides has been higher among small–marginal farmers moving from subsistence agriculture to the high value crops, though often with inappropriate technological baggage, but with a strong motivation to improve their social and economic status. They are indeed risk-taking small agricultural entrepreneurs whose success would be the basic premises for transformation of rural India towards better and equitable incomes and livelihood. To sum up, 'farmers' distress is not due to lack of agricultural growth but paradoxically due to enterprising qualities of farmers who pursue growth and even achieve it in good measure. But, drought-prone environment and non-caring policy regime turn those who bring growth into victims' (Rao & Gopalappa 2004).

Recognizing growing disparities between the agricultural and non-agricultural sectors and deterioration of the quality of the public services in rural India, some have suggested a radically different approach to make the farm sector to improve its growth performance (Vaidyanathan 2006). It is a cruel paradox that the country is agriculturally self-sufficient, and the policy makers have designs and dreams about high export growth of agricultural commodities including foodgrains, but farmers who are the architects of these surpluses are allowed to die out of distress. What is needed is a caring policy but what exists is exposure to predatory market forces instead. There is increasing evidence that there cannot be rural development, even in relatively prosperous regions such as Andhra Pradesh and Punjab, without high agricultural growth. Nor is there any instance in the world of dryland farmers moving to high productivity agriculture in the face of gross exposure to the volatile market forces. There is no instance of small–marginal farmers earning adequate livelihood neither without appropriate social security and economic support nor without succour provided by supplementary non-farm employment. Small–marginal farmers in dry regions are the most vulnerable but least cared for in the economic reforms framework. It is the policy of neglect and misguided technology path that has been forcing these farmers to shoulder all the costs and risks of high investment, including land and groundwater resource development with borrowed capital at usurious interest rates. They have lifted agriculture to an extent to a relatively better productivity but they cannot afford shouldering the high costs of risks and uncertainties without adequate state support.

These costs are the costs of transition of agriculture in vast areas of the country from subsistence farming to commercial farming, aiming towards higher productivity. These costs are necessarily social costs, which should not be compounded on to the shoulders of the distressed peasantry. The state has to own the responsibility for these social costs of investment in the development of land and water resources (including groundwater in potential areas), provision of adequate economic support by way of institutional credit, extension, supply of quality inputs, and remunerative prices as well as social sector support of ensuring quality education and health facilities in the countryside. There are a number of sensible suggestions for the revival of the agrarian conditions, the most recent of which come from the *Report of the Expert Group on Agricultural Indebtedness* (Government of India 2007) which calls for agricultural technology that involves a judicious combination of traditional wisdom and appropriate results of modern science, institutional arrangements with critical role to collectives of farmers at the grassroot level, reforms that would improve equity in access to land and water resources, and interventions that would mitigate the risks. There is incontrovertible evidence that agricultural growth driven by improved productivity of small–marginal farmers would result in much more equitable distribution of income and augmentation of effective demand with its spread effects on the non-farm sector and would be more sustainable as well. The essential condition is the need for policy shift from the mindless neoliberal market-centred reforms to the building of economic and social support systems to make small–marginal farming, especially in dry regions, viable and sustainable, and to ensure these farmers against exposure to distress due to vagaries of weather and vulnerability to domestic and global market forces.

NOTES

* An earlier version of this paper 'Economic Reforms, Agricultural Crisis and Rural Stress in India' by the first author formed part of a larger project directed by Andrea Cornia of the the University of Florence, Italy. We are grateful to him for his keen interest in the subject and very helpful comments and to Jayati Ghosh who was instrumental in involving the author in the project. The paper benefited from extensive comments by Barbara Harris-White, R. Radhakrishna, V.M. Rao, and S.R. Hashim who read the earlier version of this paper and sustained the work on the revision by their comments and encouragement. It also formed a substantial part of the Prof.

B. Janardhan Rao Memorial Lecture held on 27 February 2006 at Kakatiya University, Warangal, and was presented in other seminars where it benefited from the comments of participants, of whom special mention may be made of E. Revathi, S. Galab, and Wendy Olsen. Some of the discussions also find mention in Reddy and Mishra (2008).

1. Attempt to suicide is considered a criminal act as per the Indian Penal Code (IPC) 309. There have been court rulings calling for a humane perspective, but without legislative backing the statute remains.

REFERENCES

Action for Welfare and Awakening in Rural Environment (AWARE) (1998), *Farmer's Suicides in Andhra Pradesh*, Hyderabad: AWARE.

Acharya, S. S. (2004), 'Fertilizer Subsidy in Indian Agriculture: Some Issues', in B. Doria and T. Jullien (eds), *Agricultural Incentives in India: Past Trends and Perspective Paths Towards Sustainable Development*, New Delhi: Manohar and Centre de Sciences Humaines, pp. 67–82.

Assadi, M. (1998), 'Karnataka: Farmers' Suicides—Signs of Distress in Rural Economy', *Economic and Political Weekly*, Vol. 33, No. 14, 4 April, pp. 747–8.

Bhalla, Sheila (2005), 'India's Rural Economy: Issues and Evidence', Working Paper No. 25, Institute for Human Development, New Delhi.

Bhalla, G. S. (2006), *Condition of Indian Peasantry*, New Delhi: National Book Trust, India.

——, (2007), *Indian Agriculture since Independence*, New Delhi: National Book Trust, India.

Bhalla, G. S. and G. Singh (2001), *Indian Agriculture: Four Decades of Development*, New Delhi: Sage Publications.

Bhushan, S. and T. P. Reddy (2004), 'A Moving into Poverty Syndrome: Debt and Differentiation in Small Farm Economies: A Casual Study of Farmers' Suicides in AP', mimeo, Poverty and Social Analysis Monitoring Unit (PSAMU-SERP), Hyderabad, November.

Byres, T. J. (ed.) (1994), *The State and Development Planning in India*, New Delhi: Oxford University Press.

Central Statistical Organisation (CSO), *National Accounts Statistics*, various years.

Centre for Environmental Studies (CES) (1998), 'Gathering Agrarian Crisis: Farmers' Suicides in Warangal District (AP): A Citizen's Report', Hanamkonda (AP): CES.

Chand, R. (2006), 'India's Agricultural Challenges and Their Implications for Growth and Equity', Paper presented at Silver Jubilee Seminar on

Perspectives on Equitable Development: International Experience and What can India Learn? Centre for Economic and Social Studies, Hyderabad.

Cornia, G. A., R. Jolly, and F. Stewart (eds) (1987), *Adjustment with Human Face*, 2 Vols, Clarendon: Oxford University Press.

Deshpande, R. S. and N. Prabhu (2005), 'Farmers' Distress: Proof Beyond Question', *Economic and Political Weekly*, Vol. 40, No. 44–45, 29 October, pp. 4663–5.

Dorin, B. and T. Jullien (2004), 'The Product-Specificity of Indian Input Subsidies: Scope and Effects on Equity and Competitiveness', in B. Dorin and T. Jullien (eds), *Agricultural Incentives in India: Past Trends and Perspective Paths Towards Sustainable Development*, New Delhi: Manohar and Centre de Sciences Humains, pp. 151–94.

Durkheim, E. (2002) [1897], *Suicide* (translated from French by J.A. Spaulding and G. Simpson), London and New York: Routledge Classics.

Government of India (2002), *Ninth Five Year Plan 1997–2002*, New Delhi: Planning Commission.

—— (2005a), *Mid-term Appraisal of the Tenth Five Year Plan*, New Delhi: Planning Commission.

—— (2005b), *Economic Survey 2004–05*, New Delhi: Ministry of Finance.

—— (2007), *Report of the Expert Group on Agricultural Indebtedness*, New Delhi: Ministry of Finance, July.

Gulati, A. and S. Narayanan (2003), *The Subsidy Syndrome of Indian Agriculture*, New Delhi: Oxford University Press.

Gupta, S. (2005), 'Crisis of Indian Agriculture under Neoliberal Policy', Working Paper No. 24, Institute for Human Development, New Delhi.

Harris-White, B. and S. Janakaranjan (2004), *Rural India Facing the 21st Century: Essays on Long Term Village Change and Recent Development Policy*, London: Anthem Press.

International Labour Organization (ILO) (2005), *World Employment Report 2004–05: Employment, Productivity and Poverty Reduction*, Geneva: International Labour Organization.

Joshi, P. K., A. Gulati, and R. Cummings Jr (eds) (2007), *Agricultural Diversification and Small Holders in South Asia*, New Delhi: Academic Foundation.

Kumar, P. (2005), 'Empowering the Small Farmers Towards a Food Secure India', in R. Chand (ed.), *India's Agricultural Challenges: Reflections on Policy, Technology and Other Issues*, New Delhi: Centad, pp. 197–225.

Mishra, S. (2006a), *Suicide of Farmers in Maharashtra*, http://www.igidr.ac.in/suicide/suicide.htm, Report Submitted to the Government of Maharashtra, Mumbai: Indira Gandhi Institute of Development Research.

—— (2006b), 'Farmers' Suicides in Maharashtra', *Economic and Political Weekly*, Vol. 41, No. 16, 22 April, pp. 1538–45.

—— (2006c), 'Suicide Mortality Rates Across States of India, 1975–2001: A Statistical Note', *Economic and Political Weekly*, Vol. 41, No. 16, 22 April, pp. 1566–9.

—— (2006d), 'Suicides in India: Some Observations', in K.S. Bhat and S. Vijaya Kumar (eds), *Undeserved Death: A Study on Suicide of Farmers in Andhra Pradesh (2000–2005)*, New Delhi: Allied Publishers, pp. 93–113.

Mohan Rao, R. M. (2004), *Suicides Among Farmers—A Study of Cotton Growers*, New Delhi: Concept Publishing Company.

Mohanakumar, S. and R. K. Sharma (2006), 'Analysis of Farmer Suicides in Kerala', *Economic and Political Weekly*, Vol. 41, No. 16, 22 April, pp. 1553–8.

Mohanty, B. B. (2005), '"We are Like the Living Dead": Farmer Suicides in Maharashtra, Western India', *The Journal of Peasant Studies*, Vol. 32, No. 2, April, pp. 243–76.

Nair, K. N. and V. Menon (2005), 'Reforming Agriculture in a Globalizing World—The Road Ahead for Kerala', IP6 Working Paper No. 3, NCCR—North South (Swiss National Science Foundation, Berne).

Narendranath, G., U. Shankari, and K. R. Reddy (2005), 'To Free or Not to Free Power: Understanding the Context of Free Power to Agriculture', *Economic and Political Weekly*, Vol. 40, No. 53, 31 December, pp. 5561–70.

National Crime Records Bureau (NCRB) (Various Years), *Accidental Deaths and Suicides in India*, New Delhi: Ministry of Home Affairs, Government of India.

National Sample Survey Organisation (NSSO) (Various Rounds), *Employment and Unemployment Situation in India*, New Delhi: Ministry of Statistics and Programme Implementation, Government of India,

—— (2005), *Situation Assessment Survey of Farmers: Indebtedness of Farmer Households*, NSS 59th Round (January–December 2003), Report No. 498 (59/33/1), New Delhi: Ministry of Statistics and Programme Implementation, Government of India.

—— (Various Rounds), *Household Consumer Expenditure and Employment Situation in India*.

—— (Various Rounds), *Some Aspects of Operational Land Holdings in India*.

Pingali, P. (2005), 'Green Revolution to Gene Revolution', in *Vistas in Agricultural Marketing (1997–2005)*, Vol. 2, Nagpur: India Society of Agricultural Marketing, pp.176–97.

Ramchandran, V. K. and M. Swaminathan (eds) (2005), *Financial Liberalization and Rural Credit in India*, New Delhi: Tulika Books.

Rao, C. H. H. (1975), *Technological Change and Distribution of Gains in Indian Agriculture*, New Delhi: Macmillan.

Rao, P. N. and K. C. Suri (2006), 'Dimensions of Agrarian Distress in Andhra Pradesh', *Economic and Political Weekly*, Vol. 41, No. 16, 22 April, pp. 1546–52.

Rao, P. S. M. (2004a), 'Weaker Sections' Rural Credit in India: A Post-Reform Scenario', mimeo.

—— (2004b), 'Growing Rural Indebtedness and Increasing Institutional Apathy', mimeo.

Rao, V. M. (1992), 'Land Reform Experiences: Perspective for Strategy and Programmes', *Economic and Political Weekly*, Vol. 27, No. 26, 27 June, pp. A50–A64.

—— (2004), *State of the Indian Farmer: A Millennium Study—Rainfed Agriculture*, Vol. 10, New Delhi: Academic Foundation.

Rao, V. M. and D. V. Gopalappa (2004), 'Agricultural Growth and Farmer Distress: Tentative Perspectives from Karnataka', *Economic and Political Weekly*, Vol. 39, No. 52, 25 December, pp. 5591–8.

Ray, S. (ed.) (2007), *Handbook of Agriculture in India*, New Delhi: Oxford University Press.

Reddy, D. N. (2006a), 'Changes in Agrarian Structure and Agricultural Technology: Is Peasant Farming Sustainable under Institutional Retrogression', in R. Radhakrishna, S. K. Rao, S. Mahendra Dev and K. Subba Rao (eds), *India in a Globalising World: Some Aspects of Macroeconomy, Agriculture and Poverty, Essays in Honour of C.H. Hanumantha Rao*, New Delhi: Academic Press, pp. 409–28.

—— (2006b), 'Economic Reforms, Agrarian Crisis and Rural Distress', 4th Annual Professor B. Janardhan Rao Memorial Lecture, Warangal, Telengana: Professor B. Janardhan Rao Memorial Foundation.

Reddy, D. N. and S. Mishra (2008), 'Crisis in Agriculture and Rural Distress in Post-reform India', in R. Radhakrishna (ed.) *India Development Report 2008*, New Delhi: Oxford University Press, pp. 40–53.

Rythu Sahaya Committee (RSC) (1998), 'Farmers' Suicides in Andhra Pradesh: Report of the People's Tribunal', Hyderabad: RSC, July.

Sen, A. (2003), 'Globalisation, Growth and Inequality in South Asia: The Evidence from Rural India', in J. Ghosh and C. P. Chandrasekhar (eds), *Work and Well-being in the Age of Finance*, New Delhi: Tulika Books, pp. 469–507.

Sen, A. and M. S. Bhatia (2004), *State of the Indian Farmer: A Millennium Study—Cost of Cultivation and Farm Income*, Vol. 14, New Delhi: Academic Foundation.

Shetty, S. L. (2006), 'Monetary Policy and Financial Sector Liberalization', in *Macroeconomics of Poverty Reduction: India Case Study*, Report submitted to United Nations Development Programme, Mumbai: Indira Gandhi Institute of Development Research, April.

Shiva, V., A. H. Jafri, A. Emani, and M. Pande (2000), *Seeds of Suicide: The Ecological and Human Costs of Globalisation in Agriculture*, New Delhi: Research Foundation for Science, Technology and Ecology.

Vaidyanathan, A. (2006), 'Farmers' Suicides and the Agrarian Crisis', *Economic and Political Weekly*, Vol. 41, No. 38, 23 September, pp. 4009–13.

Vakulabharanam, V. (2005), 'Growth and Distress in a South Indian Peasant Economy During the Era of Economic Liberalization', *Journal of Development Studies*, Vol. 41, No. 6, August, pp. 971–97.

Vakulabharanam, V. and S. Motiram (2007), 'Political Economy of Agrarian Distress in India Since the 1990s', Paper presented at Columbia–LSE–New School Conference on 'Great Transformation: Understanding India's New Political Economy', Columbia University, New York, September.

Vasavi, A. R. (1999), 'Agrarian Distress in Bidar: Market, State and Suicides', *Economic and Political Weekly*, Vol. 34, No. 32, 7 August, pp. 2263–8.

Venugopal, P. (2004), *State of the Indian Farmer: A Millennium Study—Input Management*, Vol. 8, New Delhi: Academic Foundation.

Vidyasagar, R. and K. Suman Chandra (2004), *Farmers' Suicides in Andhra Pradesh and Karnataka*, Hyderabad: National Institute of Rural Development.

2

Capital Formation in Indian Agriculture
National and State Level Analysis

RAMESH CHAND

INTRODUCTION

Although economic reforms initiated in 1991 have put the Indian economy on a higher growth trajectory, the agriculture sector, which accounted for more than 30 per cent of the Gross Domestic Product (GDP) at the start of the reforms, failed to keep pace with the growth of the non-agriculture sector (Chand 2005). On the contrary, it was a sharp deceleration in the agricultural growth rate that was witnessed after the mid–1990s.[1] The trend of growth in the later period was much less than the rate of growth of rural population. The implication is that per capita income in agriculture is declining, resulting in rising rural and agricultural distress in the country. While agricultural growth *per se* is affected by a large number of factors such as irrigation, fertilizers, area, cropping intensity, cropping pattern, and technology (Chand et al. 2007), the overall condition of not only agricultural production but also agriculture as a remunerative pursuit of livelihood depends on the behaviour of prices of inputs and output, state support systems, and public investment. Of these factors, public sector capital formation in agriculture is of critical importance. There is renewed interest in this issue in the wake of sharp deceleration in the growth rate of Indian agriculture and serious distress being faced by Indian farmers in several parts of the country, which is believed to be related to poor growth in agricultural income.

A large number of studies have shown that since 1980–1 public sector capital formation in Indian agriculture has continuously followed a declining trend, with some short breaks (Rath 1989; Shetty 1990; Hanumantha Rao 1994; Alagh 1994; Chand 2001). Almost the entire literature on trends in capital formation and its implications is focused at the national level and very few studies have looked at state level trends in public investment in agriculture. The primary reason for this scant attention to the state level scenario has been non-availability of state level data on capital formation. This data at the state level has been reported in a very scanty fashion, that too only for a few years. In contrast to state level data, estimates of private as well as public investment in agriculture at the country level are available on an annual basis from the Central Statistical Organisation (CSO) since 1950–1. While country level series are important to assess the status of capital formation at the aggregate level, they do not reveal variations in agricultural investment across states, which is vital for determining the state in which more urgent action is needed.

This chapter examines the trends in the national level public and private investment and state level public investment in agriculture and attempts to explore the factors that affect growth in various types of investment. It also analyses the implications of the underlying trends in public and private investment on output growth. Due to non-availability of state level data on capital formation, the chapter takes recourse to data on capital expenditure on various heads in agriculture proper and in related heads.

Capital Formation at National Level

The trend in capital formation in agriculture is presented in Figure 2.1 and Table 2.1. During 1970–1 to 1974–5 there was no clear growth in public sector and private sector investments. After this both public as well as private sector investment started increasing rapidly. The growth in public sector investment continued upto the year 1980–1 when India invested Rs 7,358 crore in agriculture at 1993–4 prices, corresponding to 3.78 per cent of agricultural GDP at current prices. This turned out to be the historic peak level of investment and even

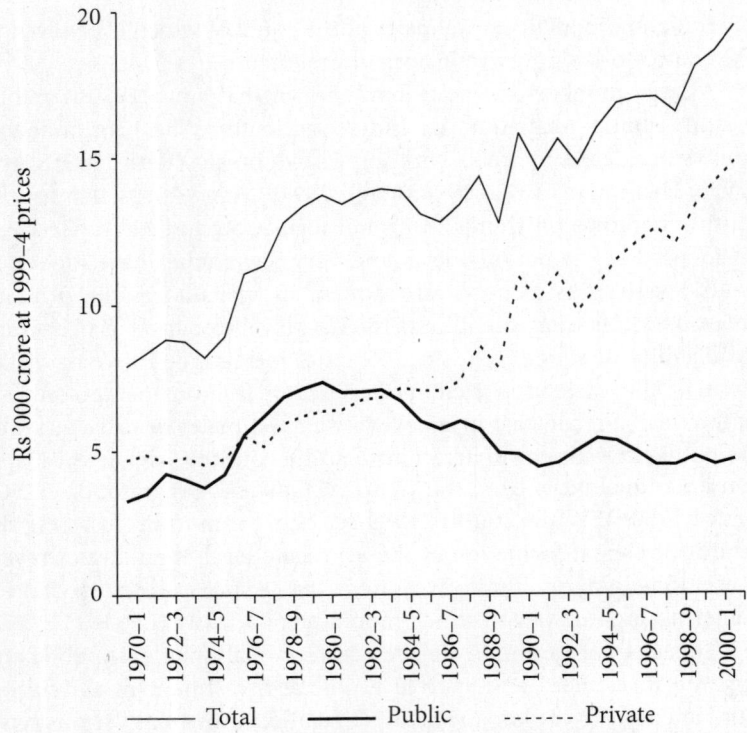

Figure 2.1: Gross Fixed Capital Formation in Agriculture: 1970–1 to 2001–2

Source: CSO, *National Accounts Statistics*, various years.

after more than two decades this level could not be attained. On the contrary, the level of public investment declined year after year and the recent level of public investment is less than three-fourth of the level attained during 1980–1.

The year 1980–1 also marked the beginning of a significant change in the composition of total investment in agriculture. Till that year the share of public and private investment did not differ much from each other and both the series show a similar direction of trend. After 1980–1, public investment in agriculture followed a declining trend but private investment in agriculture showed a robust increase year after year. Initially, when public investment in agriculture started declining after 1980–1, it caused some adverse impact on total investment in

Table 2.1
Public, Private, and Total Investment in Agriculture

(Rs crore at 1993–94 prices)

Year	Total	Public	Private	Year	Total	Public	Private
1970–1	7,902	3,216	4,686	1987–8	13,375	6,004	7,371
1971–2	8,349	3,478	4,871	1988–9	14,335	5,733	8,602
1972–3	8,831	4,212	4,619	1989–90	12,728	4,911	7,817
1973–4	8,760	3,983	4,777	1990–1	15,805	4,871	10,934
1974–5	8,212	3,691	4,521	1991–2	14,546	4,400	10,146
1975–6	8,924	4,185	4,739	1992–3	15,610	4,549	11,061
1976–7	11,066	5,566	5,500	1993–4	14,749	4,996	9,753
1977–8	11,347	6,191	5,156	1994–5	15,978	5,406	10,572
1978–9	12,780	6,848	5,932	1995–6	16,824	5,318	11,506
1979–80	13,344	7,141	6,203	1996–7	17,009	4,942	12,067
1980–1	13,721	7,358	6,363	1997–8	17,035	4,467	12,568
1981–2	13,407	6,998	6,409	1998–9	16,516	4,459	12,057
1982–3	13,766	7,020	6,746	1999–2000	18,082	4,764	13,318
1983–4	13,926	7,089	6,837	2000–1	18,602	4,471	14,131
1984–5	13,846	6,699	7,147	2001–2	19,402	5,489	13,913
1985–6	13,061	6,005	7,056	2002–3	19,280	4,788	14,492
1986–7	12,789	5,738	7,051				

Source: CSO, National Accounts Statistics, various years.

agriculture, but subsequently the sharp increase in private investment resulted in an increase in total investment as well.

Figure 2.1 shows three distinct phases in the movement of agricultural investment. These are:

(a) 1974–5 to 1980–1: This phase witnessed rapid rise in public investment, moderate increase in private investment and fairly high growth in total investment.

(b) 1980–1 to 1989–90: Decline in public investment, moderate increase in private investment, and stagnation in total investment.

(c) 1989–90 to 2001–2: Stagnation in public investment, high growth in private investment, and moderate growth in total investment.

Table 2.2

Growth Rate of Public, Private, and Total Investments in Agriculture
(at 1993–4 prices)

(per cent)

Period	Total investment	Public investment	Private investment
1974–80	8.90	12.83	5.17
1981–90	−0.25	−3.48	2.67
1991–2003	2.60	0.05	3.71

Source: Based on Table 2.1.

The growth rate in total, public, and private sector investment during the above-mentioned phases is shown in Table 2.2. During 1974–80, public investment increased annually by 12.83 per cent and private investment increased by more than 5 per cent per annum. This adds up to 8.9 per cent annual growth in total investment in agriculture. During the 1980s, public investment in real terms declined annually by 3.48 per cent. However, private investment in this period kept increasing, but its growth rate became almost half compared to the growth rate in 1974–80. Total investment showed a trend growth of −0.25 per cent in this period. Since 1991, the level of public investment remained stagnant and fluctuated around Rs 4840 crore. Private investment showed a growth of 3.71 per cent per annum in this period, which pulled up total investment to a positive trend of 2.6 per cent in spite of virtual stagnation in the growth of public investment. However, private investment as a share of agricultural GDP did not show any improvement.

Another way to examine the progress in capital formation in agriculture is by looking at the ratio of gross fixed capital formation (GFCF) in sectoral GDP. The level of capital formation was around 5 per cent during the early 1970s. After this it increased sharply to around 8 per cent by 1979–80. This increase was followed by a decline and during the whole of the 1980s the total capital formation in agriculture hovered around 7.25 per cent. During 1990–1 to 1998–9, the level of capital formation further declined to about 6.5 per cent. There was some recovery after 1998–9, which raised the level of capital formation close to 7 per cent towards 2003–4, but still lower than what it was at the end of the 1970s.

Public Sector Capital Formation and Subsidies in Agriculture

It is interesting to observe that the decline in public sector investment after 1981–5 coincided with an increase in subsidies (Table 2.3, Figures 2.2 and 2.3). During the five years from 1980–1 to 1984–5, the level of public investment was 3.51 per cent of agricultural GDP, while subsidies were at 4.01 per cent of sectoral GDP. During 1985–6 to 1989–90, the magnitude of public investment declined to 2.96 per cent, whereas the level of subsidies increased to 4.96 per cent. This trend is continuing since then.

The increase in subsidies alongside the decrease in public investment is sometimes seen to be caused by resource transfer from the capital account to the revenue account to meet the rising bill of subsidies. According to Gulati and Narayanan (2003, pp. 202–3), burgeoning subsidies compete for scarce resources and impinge upon the government's ability to invest in key areas. They assert that increasing subsidies crowd out public investment, and this is a dangerous trend as declining public investment endangers even food security and could put a brake on the growth process. The study also opines that declining subsidies would accommodate and enable a rise in meaningful investment in crucial areas such as irrigation and infrastructure.

Table 2.3

Trend in Investment and Subsidies in Agriculture Expressed as per cent of Agricultural GDP

Period	Total investment	Public investment	Private investment	Subsidy	Public investment and subsidy
1971–5	4.99	2.04	2.95	1.21	3.25
1976–80	7.07	3.39	3.68	2.95	6.34
1981–5	7.28	3.51	3.77	4.01	7.52
1986–90	7.05	2.96	4.09	4.96	7.92
1991–5	6.68	2.09	4.60	5.17	7.26
1996–2000	6.36	1.91	4.45	5.67	7.57
2001–3	6.75	1.41	3.60	7.42	8.84

Sources: CSO, *National Account Statistics*, various years; Government of India, *Economic Survey*, various years.

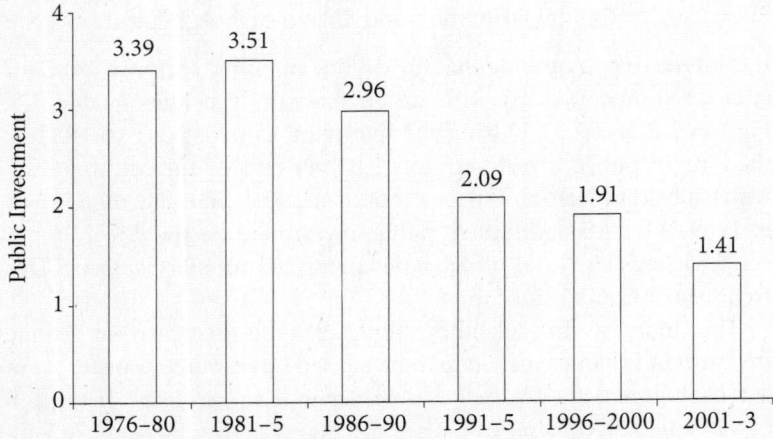

Figure 2.2: Public Investment as per cent of Agricultural GDP

Sources: CSO, *National Account Statistics*, various years; Government of India, *Economic Survey*, various years.

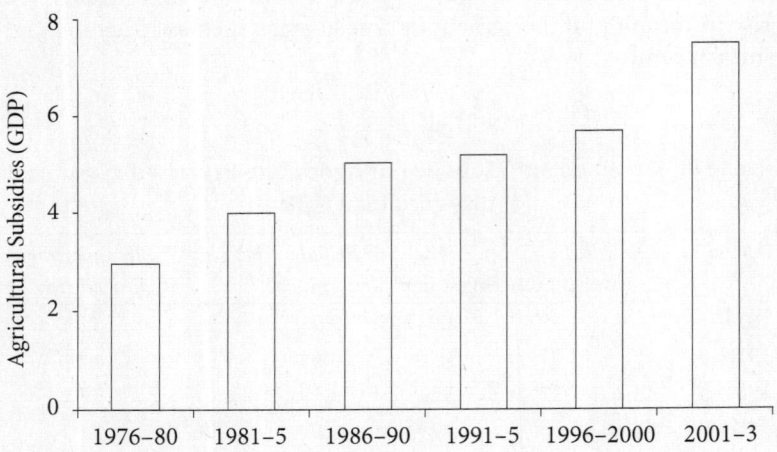

Figure 2.3: Agricultural Subsidies as per cent of Agricultural GDP

Sources: CSO, *National Account Statistics*, various years; Government of India, *Economic Survey*, various years.

Mishra and Chand (1995) find a political economy explanation for decline in public sector capital formation in agriculture. According to them, the Green Revolution during the decade of the 1970s galvanized the farmers into politically conscious, interest-seeking groups and at the national level as a 'class for itself' *vis-à-vis* the government policies. Towards the closing years of the 1970s, the state virtually lost its autonomy and agricultural policies increasingly came to be decided by farmers' interest-groups. During the 1980s, farming group-interest did prevail upon the determination of allocation of public funds committed for agriculture, as between expenditure on current and capital account. These interest groups pushed up the share of current account expenditure, which yielded input subsidies and higher support prices. Rapid increase in the share of current account expenditure to meet their demand for production subsidies and priority to finance private sector capital formation did not leave much room for the state to increase public sector capital formation. Mishra and Chand explained that these political factors were responsible for the public expenditure on agriculture taking increasingly the form of current account expenditure typified by food and fertilizer subsidies at the cost of public sector capital formation in agriculture during the 1980s. Thus, the real cause for decline in public investment in real terms during the 1980s lies in the politics of the state's agricultural policies. Mishra and Chand also noted the role of extraneous forces that emerged during the 1980s and restricted whatever autonomous choice the state had remaining for pushing up public capital formation in agriculture, which has largely been in the major and medium irrigation systems. First was the escalation of the per hectare cost of these systems compared to the minor irrigation sources, largely in the private sector. The second was the forceful rise of the environmentalist movement, domestic as well as foreign, against these systems during the 1980s. By obstructing the construction of these systems, the domestic environmentalists incredibly added, and continue to add, to their gestation lags and thereby to the cost. Further, the endemic problem of interstate disputes about water sharing became more severe.

The issue of tradeoffs between public investment and input subsidies was empirically examined by Chand and Kumar (2004) who found that the increase in subsidies and decline in revenue contribution of the agriculture sector have caused adverse impact on public sector capital formation. The paper estimated that one rupee going into public investment is ten times more productive over its lifetime as compared to the contribution of subsidies to output growth. Thus, there exists a very strong case for reducing agricultural subsidies and increasing public investment in agriculture.

Coming back to the trends in public investment and input subsidies after 1980, it is observed that initially (during 1984–5 to 1989–90) the increase in subsidies was more than the decline in public investment, and there was an increase in ratio of the sum of public investment and input subsidies in agricultural GDP till 1990. In the next ten years, that is, in the decade of the 1990s, even the increase in subsidies in real terms fell short of the decrease in public investment, and the ratio of subsidies plus investment to agricultural GDP fell from 7.92 per cent during 1986–90 to 7.42 for the decade of 1990s as a whole. This shows that even total support to agriculture through investment and subsidies during 1991–2000 declined as compared to 1981–90. Thus, during the first decade of economic reforms not only public investment in agriculture reached an all-time low of below 2 per cent of agricultural GDP, even the sum of subsidies and public investment as a proportion of agricultural GDP showed a decline.

Diversion of resources from public investment to subsidies has two more adverse consequences. First, subsidies in several cases are doing more harm than benefit through indiscriminate use of water resources, degradation of land, and imbalances in the use of plant nutrients. Second, subsidies provided by the Central government such as fertilizer subsidies, are highly skewed as the use of such inputs is very low in low-productivity states and quite high in the high-productivity states. A rise in public investment on the contrary can act a corrective to regional disparities by augmenting the productivity base in states that are agriculturally backward.

EXPENDITURE APPROACH TO MEASURE PUBLIC INVESTMENT AT STATE LEVEL

Due to non-availability of state level data on capital formation, the chapter takes recourse to budgetary data of the states on capital expenditure under various heads in agriculture proper and in related heads. These include four major heads, namely (a) agriculture and allied, (b) rural development, (c) special area programmes, and (d) major and medium irrigation and flood control. The first category includes 11 items, namely (i) crop husbandry, (ii) soil and water conservation, (iii) animal husbandry, (iv) dairy development, (v) fisheries, (vi) forestry and wild life, (vii) plantations, (viii) food storage and warehouses, (ix) agricultural research and education, (x) cooperation, and (xi) others. Out of these items, capital expenditure on forestry and wildlife was not included in the series, as most of that is not related to agriculture sector.

Capital formation results from capital expenditure. Therefore, there is a close correspondence between capital expenditure and capital formation. In the absence of data on capital formation, the data on capital expenditure can be used as a proxy for capital formation. This is what has been done here to examine public investment or capital formation in agriculture at the state level. Due to very large fluctuations in annual data, changes in capital expenditure at the state level have been examined by taking an average of selected periods. The estimates for the Eighth and Ninth Five Year Plans and for the first three years of the Tenth Five Year Plan are presented in Tables 2.4 and 2.5. Total real capital expenditure (RCE) on agriculture and related heads (at 1993–4 prices), for all states, shows a very small increase between the Eighth and Ninth Five-Year Plans. Annual allocations between the two plans increased from Rs 6250 crore to Rs 6572 crore. This amounts to a mere 5.2 per cent increase in a five-year period. There was a substantial increase in capital expenditure on agriculture during the Tenth Plan. Average capital expenditure during the first three years was Rs 10,440 crore, which was 59 per cent higher than the allocation for public investment during the Ninth Five Year Plan.

There is considerable variation in state level trends in funds allocated for capital formation. Andhra Pradesh, Assam, Bihar, Gujarat, Karnataka, Madhya Pradesh, Maharashtra, and Uttar Pradesh show a sharp increase in public sector capital expenditure in agriculture during the Ninth and Tenth Five-Year Plans. In contrast to this, there was a sharp decline in the states of Kerala, Haryana, Orissa, Punjab, and West Bengal; all these states except Orissa are agriculturally well developed, and rank among the top states in agricultural productivity (Table 2.5).

Table 2.4

Real Capital Expenditure on Agriculture and Allied Heads
(at 1993–4 prices)

(Rs crore)

State	Eighth Plan (1993–4 to 1996–7)	Ninth Plan (1997–8 to 2001–2)	Tenth Plan (2002–3 to 2004–5)
Andhra Pradesh	622	701	1,231
Assam	105	127	369
Bihar*	217	1,181	1,545
Chhattisgarh		111	275
Gujarat	540	767	853
Haryana	127	198	98
Himachal Pradesh	23	33	48
Jammu and Kashmir	103	91	111
Jharkhand		685	670
Karnataka	681	831	1,304
Kerala	149	98	75
Madhya Pradesh*	444	598	1,444
Maharashtra	1,276	970	2,616
Orissa	221	306	177
Punjab	277	351	202
Rajasthan	489	482	564
Tamil Nadu	87	201	234
Uttarakhand		51	68
Uttar Pradesh*	513	716	1,070
West Bengal	183	142	110
Arunachal Pradesh	13	18	30
Manipur	39	33	43
Meghalaya	14	14	25
Mizoram	17	15	29
Nagaland	14	16	30
Sikkim	1	5	7
Tripura	28	49	71
Goa	28	24	28
All States	6,250	6,576	10,443

Note: *Capital expenditure in states like Bihar, Madhya Pradesh, and Uttar Pradesh, including separated states, namely Jharkhand, Chhattishgarh, and Uttarakhand.
Source: Reserve Bank of India, 'Finances of State Governments', *Reserve Bank of India Bulletin*, various issues.

Table 2.5

Capital Expenditure on Agriculture and Related Heads in Relation to NSDP and Area

(Rupees)

State	RCE/ha net sown area			NSDP/ha.	Capital expenditure as % of NSDP agri.		
	1993–4 to 1996–7	1997–8 to 2001–2	2002–3 to 2004–5	2002–3 to 2004–5	1993–4 to 1996–7	1997–8 to 2001–2	2002–3 to 2004–5
Andhra Pradesh	590	660	1,193	37,383	3.51	3.60	6.26
Assam	383	466	1,350	40,374	1.96	2.27	6.57
Bihar*	297	666	1,174	32,921	1.68	3.25	4.98
Chhattisgarh		93	574	11,167	–	4.05	10.85
Gujarat	563	799	903	23,711	4.49	6.90	6.68
Haryana	357	556	279	46,304	1.49	2.30	1.01
Himachal Pradesh	409	602	868	66,354	1.87	2.17	2.57
Jammu and Kashmir	1,408	1,232	1,478	66,274	5.17	4.58	5.27
Jharkhand		775	3,787	42,817	–	13.47	12.23
Karnataka	645	811	1,300	19,877	5.29	5.21	10.54
Kerala	660	435	341	55,221	2.28	1.94	1.62
Madhya Pradesh*	225	248	602	14,109	2.56	2.45	5.78
Maharashtra	710	548	1,484	19,175	6.40	5.15	15.06
Orissa	358	509	304	23,786	3.39	5.00	3.29

(Contd.)

(Table 2.5 contd.)

State	RCE/Ha net sown area			NSDP/ha.	Capital expenditure as % of NSDP agri.		
	1993–4 to 1996–7	1997–8 to 2001–2	2002–3 to 2004–5	2002–3 to 2004–5	1993–4 to 1996–7	1997–8 to 2001–2	2002–3 to 2004–5
Punjab	663	827	475	63,988	2.15	2.33	1.23
Rajasthan	293	299	356	13,037	4.00	3.53	4.77
Tamil Nadu	153	373	468	33,231	0.76	1.74	2.68
Uttaranchal		257	863	50,923	–	2.11	2.87
Uttar Pradesh*	296	378	569	38,839	1.62	1.89	2.56
West Bengal	335	260	202	70,524	1.06	0.75	0.75
Arunachal Pradesh	731	1,079	1,847	34,599	4.94	5.29	9.14
Manipur	2,803	2,285	2,759	55,842	10.59	6.92	7.53
Meghalaya	694	622	1,106	44,632	3.85	2.62	3.94
Mizoram	2,101	1,539	3,310	51,889	9.22	6.85	12.43
Nagaland	709	900	2,784	80,699	5.18	3.25	4.10
Sikkim	114	470	563	22,672	0.84	4.49	5.60
Tripura	1,005	1,751	2,524	53,376	5.73	5.70	7.45
Goa	2,014	1,714	1,958	161,727	13.78	8.52	8.59
All States	438	464	751	30,348	2.85	3.16	5.17

Notes: RCE denotes real capital expenditure, Ha denotes hectares, NSDP denotes Net State Domestic Product

* Capital expenditure in states like Bihar, Madhya Pradesh, and Uttar Pradesh, including separated states, namely Jharkhand, Chhattisgarh, and Uttarakhand.

Source: As in Table 2.4.

In order to get a comparative picture of various states, capital expenditure on agriculture was compared on per hectare basis and as per cent of agricultural Net State Domestic Product (NSDP), (Table 2.5). The average of all these states shows a small increase in real capital expenditure from Rs 438 per hectare of net sown area during the Eighth Five-Year Plan to Rs 464 during the Ninth Five-Year Plan. There was a substantial increase during the Tenth Five-Year Plan, which saw on increase in per hectare allocation to Rs 751 per hectare. Across states, the highest allocation is found in the case of Jharkhand, followed by Maharashtra. The lowest allocation is observed in the case of West Bengal. From the mid-1990s (1993–4 to 1996–7) to early 2000s (2002–3 to 2004–5), capital expenditure has shown improvement in the less developed states, except Orissa, and deterioration in the more developed states.

Per hectare capital expenditure declined and became lower than the national average in Haryana, Kerala, Punjab, and West Bengal . In three (Gujarat, Karnataka, and Maharashtra) out of nine agriculturally more underdeveloped states having per hectare agricultural NSDP lower than the national average, capital expenditure was higher than the national average. These states are now paying higher attention to improve infrastructure for agriculture. On the other hand, Chhattisgarh, Madhya Pradesh, Orissa, Rajasthan, and Sikkim allocated lower than average resources for infrastructure development. Among the other major states, Bihar and Jharkhand now seem to be allocating higher resources for capital formation whereas current Uttar Pradesh, as was the case with undivided Uttar Pradesh, lags behind the national average. Average capital expenditure on agriculture in West Bengal in recent years has been less than one-third of the all-India average. Despite the very low level of capital formation, agriculture productivity in West Bengal is at a higher level.

As a per cent of agricultural NSDP, capital expenditure on agriculture and related heads accounted for 2.85 per cent of NSDP during the Eighth Five Year Plan, and increased to 3.16 per cent during the Ninth Five Year Plan. The ratio of capital expenditure to NSDP crossed the 5 per cent level during the first three years of the Tenth Five Year Plan. Among major states, the highest allocation of resources in the recent years is observed in the case of Maharashtra. In fact, Maharashtra was maintaining this lead during the Eighth and

Ninth Five Year Plans as well and the state's allocations proportionately have been three times higher compared to the average of all states. However, it is a matter of concern that agriculture productivity in Maharashtra is less than two-thirds of the all-India average despite relatively higher allocation of resources for more than a decade. All agriculturally developed states, namely, Haryana, Kerala, Punjab, and West Bengal spend less than 2 per cent of NSDP on capital formation in agriculture. It seems these states do not see much prospect of agricultural growth through infrastructure development and what is needed is technological breakthrough.

While the low level of allocation for infrastructure does not cause much effect on growth in developed states, it certainly is a major factor in states such as Orissa and Bihar. It is interesting to note that the correlation between the level of per hectare NSDP in agriculture and real capital expenditure per hectare was –0.57. This implies that state-wise allocation of resources for infrastructure development, of late, has been paying attention to the equity aspect and unexploited potential.

CONCLUSIONS

Public sector capital formation in agriculture for the country as a whole either declined or showed stagnation for a long period of more than a quarter century after 1980. This has caused an adverse impact on growth of agriculture output in the country. The decline in public investment has been accompanied by a rise in input subsidies and this suggests a possible diversion of resources from the capital account to the current account. This has several adverse implications for the future of Indian agriculture. One, resources spent as public investment are several times more productive as compared to the resources used as input subsidies. Two, subsidies have reached a level where they are causing more harm in the form of unsustainable use of water, imbalance in the use of plant nutrients, and degradation of soils. As public investment is found to have stronger effect on the growth of agriculture in the long run, diversion of resources from investment to subsidies has a net negative effect on output. Further, a cause for concern is that during the decade of reforms (1991–2000) the sum of both input subsidies and public investment as per cent of agricultural GDP shows a decline. Another reason for the decline in public investment in agriculture after 1980 is increased reliance of public policy on prices for achieving output growth. The price incentives, which were not spread to all crops, did not lead to the expected amount of private investment in agriculture. With the beginning of reforms, public policy remained focused on industry and other non-agriculture sectors, and adequate attention was not paid to the disquieting trends setting in agriculture.

An alternative approach using capital expenditure by all states on agriculture and related heads as an indicator of public sector capital formation also reveals the weakening public investment in agriculture. This was considered more relevant as more and more resources are channeled for infrastructure development through rural development projects and special area programmes at the state level. Total capital expenditure under agriculture and related heads show sluggish growth between the Eighth and Ninth Five Year Plans. However, the first three years of the Tenth Five Year Plan, for which data are available till now, show some encouraging trends. First, the annual allocation of resources for infrastructure development in agriculture during 2002–3 to 2004–5 was about 60 per cent higher than the annual allocation during 1996–7 to 2001–2, that is, the Ninth Five Year Plan. Second, resource allocation has been much higher in most of the agriculturally highly underdeveloped states as compared to highly developed states. It is pertinent to mention, however, that Chhattisgarh, Madhya Pradesh, Orissa, and Rajasthan continue to lag behind the average of all states even during the Tenth Five Year Plan. Finally, the resource allocation for capital formation in agriculture during the Tenth Five Year Plan offers a ray of optimism to reverse the declining trend in public sector investment through the allocation of more resources to the agriculturally laggard regions. This augurs well for growth as well as equity in agriculture.

NOTE

1. The agricultural growth rate has fallen from 3.2 per cent during the period 1980 to 1996–7, to a trend average of 1.5 per cent subsequently.

REFERENCES

Alagh, Y. K. (1994), 'Macro Policies for Indian Agriculture', in G. S. Bhalla (ed.), *Economic Liberalisation and Indian Agriculture*, New Delhi: Institute for Studies in Industrial Development, pp. 23–54.

Central Statistical Organisation, *National Accounts Statistics*, various years.

Chand, R. (2001), 'Emerging Trends and Issues in Public and Private Investment in Indian Agriculture: A Statewise Analysis', *Indian Journal of Agricultural Economics*, Vol. 56, No. 2, pp. 161–84.

—— (2005), 'India's National Agricultural Policy: A Critique', in R. Chand (ed.), *India's Agricultural Challenges: Reflections on Policy, Technology and Other Issues*, New Delhi: Centad, pp. 19–46.

Chand, R. and P. Kumar (2004), 'Determinants of Capital Formation and Agriculture Growth: Some New Explorations', *Economic and Political Weekly*, Vol. 39, No. 52, 25 December, pp. 5611–6.

Chand, R., S. S. Raju, and L. M. Pandey (2007), 'Growth Crisis in Agriculture: Severity and Options at National and State Levels', *Economic and Political Weekly*, Vol. 42, No. 26, 30 June, pp. 2528–34.

Government of India, *Economic Survey*, New Delhi: Ministry of Finance, various years.

Gulati, A. and S. Narayanan (2003), *The Subsidy Syndrome in Indian Agriculture*, New Delhi: Oxford University Press.

Mishra, S. N. and R. Chand, 'Private and Public Capital Formation in Indian Agriculture: Comments on Complementarity Hypothesis and Others', *Economic and Political Weekly*, Vol. 30, No. 24, 24 June, pp. A64–A79.

Rao, C. H. H. (1994), *Agricultural Growth, Rural Poverty and Environmental Degradation in India*, Delhi: Oxford University Press.

Rath, N. (1989), 'Agricultural Growth and Investment in India', *Journal of Indian School of Political Economy*, Vol. 1, No. 1, January–June, pp. 64–83.

Reserve Bank of India, *Reserve Bank of India Bulletin*, Government of India.

Shetty, S. L. (1990), 'Investment in Agriculture: Brief Review of Recent Trends', *Economic and Political Weekly*, Vol. 25, No. 7, 17 February, pp. 389–98.

3

Agricultural Credit and Indebtedness
Ground Realities and Policy Perspectives[1]

S. L. SHETTY

AN OVERVIEW

In the post-reform period, there has been loss of credit delivery momentum for vast segments of the informal sector—agriculture, small-scale industries, micro-enterprises, rural artisans, and other small borrowers. Earlier studies by the author (Shetty 2004, 2006a, 2006b) have provided extensive proof of the poor performance based on both supply and demand side indicators. The present chapter approaches the issue with a broad theoretical framework, discusses the ground reality of agricultural credit flow and indebtedness, and makes suggestions for moving forward to bridge the serious vacuum created in the institutional credit structure, particularly for farming households.

The disappointing performance of the formal banking institutions in meeting the credit needs of the informal sector calls for a more incisive probe into a number of behavioural issues at the level of borrowers, the lending institutions, and the policy-making and apex monitoring bodies. At the institutional level, it raises a number of demand and supply issues, the issues of credit absorptive capacities of real sectors, the responses of banking institutions to their own organizational weaknesses, and their responses to reform measures. This is one set of behavioural issues of institutions. At the borrowers' level, absence of firmness and laxity in loan recoveries, combined with the lack of provisions for insurance against unforeseen events and defaults, have created scant respect for repaying responsibilities, and thus created, albeit partly, an artificial situation of huge non-performing assets (NPAs). But, more intriguing issues are the behaviour at the policy-making level, and with the apex bodies such as the Reserve Bank of India (RBI), the National Bank for Agriculture and Rural Development (NABARD), and the government, which generally have sought to jettison the reasonably well-tested policy of affirmative action, the supply-leading approach to financial intermediation.[2] For reasons of political economy, the policy planners dare not give up the directed credit prescriptions for priority sectors or for agriculture within them or for weaker sections. The year 1991 was the year when Service Area Planning was getting under way. Commercial banks were drawn into village level surveys to integrate the government-sponsored programmes with the branch business plans. Knowledge of area and activity, the crucial components for risk assessment of a banker, was embedded in this exercise. But, these developments were perceived as a threat to private moneylenders. The policy decision of loan waiver cannot be seen devoid of these vested interests. It was a big setback to credit planning and targeted credit to the poor. It created such a negative behaviour among the banking personnel, the only logical alternative appeared to be withholding credit flow to rural areas. The financial sector reforms came handy in the pursuit of reversal of target lending. There has been distortion in their coverage, lackadaisical approach to monitoring, and actual decline in the number of rural bank branches. In the changed context, there is need for a fresh approach to credit delivery for the informal sector.

The supply-leading approach to the process of financial intermediation is the most viable one for supporting agriculture, the non-farm informal sector and all small borrowers. The 'credit approach' to monetary policy, as distinguished from the 'money approach', calls for calibrated interventions for better distributions of bank credit sectorally, regionally, functionally, and by size. Such an approach needs promotion of concrete institutions and instruments of credit delivery through a combination of steps for institution building and a system of checks and balances in the form of rewards and punishments.

Theoretical and Empirical Underpinnings

Though the cause and effect relationship between finance and growth has been a controversial subject in the Economics literature, in our view equally important to the role of aggregate finance in economic development is the distributional goal of finance in spreading economic opportunities regionally, sectorally, functionally, and amongst small asset-owning classes, and it is better attempted in a supply-leading strategy. Bell and Rousseau (2001) bring out the way in which financial intermediaries in India have played a leading role in influencing the economic performance; their results suggested that the financial sector, amongst other things, was not only instrumental in promoting aggregate investment and output but also in attaining finance-led industrialization. What is more, studies by Burgess and Pande (2003, 2004) and Burgess, Pande, and Wong (2004) have sought to prove that state-led branch expansion into rural unbanked locations reduced poverty across Indian states; in addition, the directed bank lending requirements were associated with increased bank borrowing among the poor, in particular low caste and tribal groups. Their studies go further and point out that while the presence of a nation-wide bank branch licensing rule between 1977 and 1990 caused banks to open relatively more branches in Indian states with lower initial financial development during the period, the reverse was true outside this period; they also find that rural branch expansion in India significantly reduced rural poverty and increased non-agricultural output.

More generally, the financial policies of the 1970s and 1980s followed the supply-leading strategy of Patrick (1966) or they resembled an endogenous growth strategy in which finance itself was seen as a crucial factor of production such as knowledge and in which the influence of institutional arrangements in regard to finance on growth rates was forcefully emphasized (Eschenbach 2004; RBI 2001). We have been emphasizing that sustained expansion in sectoral credit growth in real terms during the latter half of the 1970s and the whole of the 1980s served, *inter alia*, as an important causal factor in the acceleration of growth rates in agriculture and unregistered manufacturing in the 1980s (Shetty 2002). Similarly, the acceleration in employment growth from 1.5 per cent per annum during 1977–83 to 2.70 per cent during the period of 1983 to 1993–4, and more significantly, the non-farm employment growth in rural areas that showed an impressive performance in the 1980s (Vaidyanathan 1994; Chadha 1993; Visaria 1996), appear to have been related to better sectoral, regional, and size distribution of bank credit. With a series of bank credit-linked employment and asset-generating programmes, and other forms of state interventions for hastening the pace of rural industrialization, again largely rooted in banking sector support, the non-farm sector growth in rural areas was an obvious concomitant in the 1980s.

Contrariwise, after the financial sector reforms began in the early part of the 1990s, every banking indicator—spread of branch banking in rural and historically underbanked regions, credit-deposit ratios of these regions, credit delivery for agriculture, small-scale industries, small borrowers, and other priority areas—has received a setback. No doubt, the unprecedented growth of the banking system for two decades prior to the 1990s brought in its trail many serious infirmities in the working of the whole financial system: reduced bottom-lines, large NPAs, poor capital base and insufficiency of loan loss provisions, and organizational weaknesses leading to serious deterioration in house-keeping tasks as well as customer service. By the end of the 1980s, the post-nationalization successes had begun to wear thin. Therefore, the evolution of banking after the 1990s has reflected the enormous challenges that the public sector banks in particular have faced in cleaning up and consolidating their operations in an entirely new environment of competition and profit seeking. Apart from the onerous discipline imposed by regulatory and prudential norms as part of financial sector reforms, there has also been a sea change in the role of banks as a result of competitive opportunities thrown up in para-banking activities—merchant banking, housing finance, mutual funds, insurance and others, and above all, in the notion of universal banking and project finance. The critical questions are: How to bring about coexistence of banking reforms with societal goals of banking policies addressed at meeting the financial needs of small producers like peasant farmers? Is there an alternative but to achieve such coexistence?

DISAPPOINTING GROUND REALITY SINCE THE 1990s

The official statistics on the distribution of bank credit among different segments of the informal sector (Shetty 2004, 2006b) reveal a steady deterioration since the early 1990s. Figure 3.1 shows the deteriorating bank credit in terms of declining number of small borrower accounts in the 1990s. The shares of bank credit for agriculture, small-scale industries, and small borrowal accounts in total bank credit were at their peak levels of 17.7 per cent, 13.4 per cent, and 25.4 per cent around the end of the 1980s or early 1990s; they have steadily fallen since then and reached the lowest levels of 10.0 per cent, 4.1 per cent, and 3.7 per cent of total bank credit around 2000 or 2001 (Shetty 2006b). There were erosions in the absorptive capacities of these informal sectors, including agriculture, due to significant structural changes taking place in the Indian economy such as reductions in their GDP share. However, there is need for a more detailed work on the measurement of demand for credit from these sectors, and the nature of gap that may have grown over the years (Singh & Sagar 2004). No doubt the outstandings of bank credit for the informal sectors have risen in absolute nominal terms over the years, but one

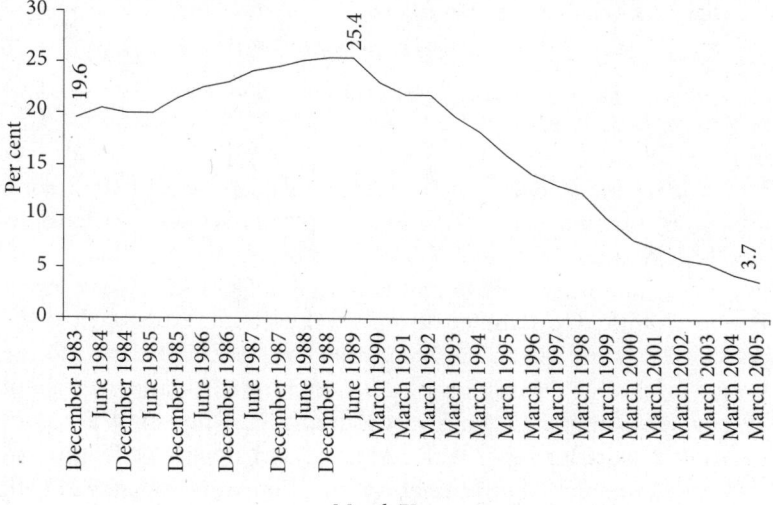

Month/Year

Figure 3.1: Percentage Share of Credit of Small Borrowal Accounts to Total Credit (by Scheduled Commercial Banks)

Note: Small Borrowal Accounts refer to credit limit of Rs 25,000 or below.
Source: Reserve Bank of India (RBI), *Basic Statistical Returns of Scheduled Commercial Banks in India*, various years.

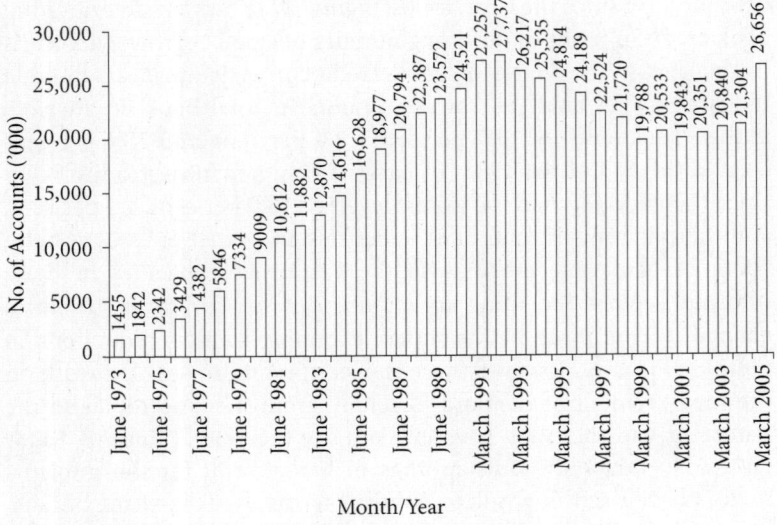

Figure 3.2: Trends in Agricultural Credit: Number of Borrowal Accounts
(for Scheduled Commercial Banks)

Source: Reserve Bank of India (RBI), *Basic Statistical Returns of Scheduled Commercial Banks in India*, various years.

incontrovertible evidence of the growing gap in the supply of bank credit for all the three categories of informal sectors—agriculture, small-scale industries, and small borrowals—lies in the drastic decline in the number of borrowal accounts from their peak levels. For agriculture, the number of borrowal accounts had touched the peak of 27.74 million in 1992, and this has steadily declined to the lowest level of 19.84 million by 2001 (Figure 3.2)—a loss of about 29 per cent. Similar declines in the number of borrowal accounts have taken place for small-scale industries (partly due to definitional changes) and, more significantly, for small borrowal categories with credit limits of Rs 25,000 or less.

Table 3.1 shows credit to agriculture from all institutional sources increased from Rs 1675 crore in 1975–6 to Rs 180,486 crore in 2005–6 and the rate of growth of credit was much higher than the growth rate of Gross Domestic Product (GDP) originating in agriculture. Despite this growth, the credit needs of agriculture

Table 3.1: Source-wise Institutional Credit Flow to Agriculture: 1975–6 to 2005–6

(Rs crore)

Agency	1975–6	1983–4	1993–4	2001–2	2003–4	2005–6
Cooperatives	1,186	2,938	10,117	23,524	26,875	39,404
	(70.80)	(56.03)	(61.34)	(37.91)	(30.89)	(21.83)
Regional rural banks	2	263	977	4854	7581	15,223
	(Neg.)	(5.02)	(5.92)	(7.82)	(8.72)	(8.43)
Scheduled commercial banks	405	1,885	5,400	33,587	52,441	125,477
	(24.16)	(35.95)	(32.74)	(54.13)	(60.29)	(69.52)
Other government agencies	82	185	–	80	84	382
	(4.90)	(3.53)	–	(0.12)	(Neg.)	(0.21)
Total credit	1,675	5,244	16,494	62,045	86,981	180,486
	(100)	(100)	(100)	(100)	(100)	(100)

Note: Neg. denotes negligible.
Source: For commercial banks from Reserve Bank of India (RBI); for cooperatives and regional rural banks from NABARD.

have not been fully met, leaving a sizeable proportion of farm households to borrow from non-institutional sources. There has been a structural shift in the institutional credit increasingly from cooperatives towards commercial banks. The share of cooperatives in the institutional credit declined from 71 per cent in 1975–6 to 22 per cent in 2005–6, while that of commercial banks increased from 24 per cent to 70 per cent (see Figure 3.3). The shift is partly due to wider deposit mobilization by the commercial banks and partly because of the failure of Primary Agricultural Cooperative Societies (PACSs) and Cooperative Central Banks (CCBs) to raise their own resources through deposit mobilization. Cooperatives have come to depend heavily on external funds from government or higher layers of cooperatives, which have their own financial constraints. Added to this, in 2003–4, more than half of PACSs reported losses (Government of India 2005). As a result, the share of cooperatives in institutional term loans declined drastically from 61.2 per cent in 1975–6 to a mere 6 per cent in 2005–6 while their share in crop loans declined from 74.9 per cent to 33.2 per cent in the same period (Table 3.2).

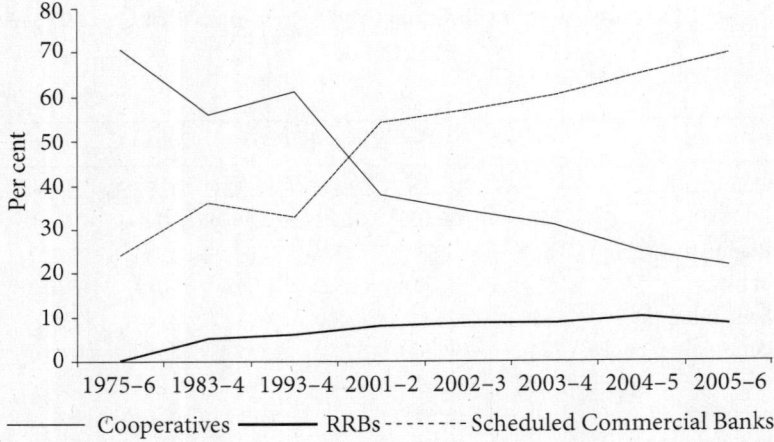

Figure 3.3: Share of Cooperatives, RRBs, and Scheduled Commercial Banks in Total Agricultural Credit: 1975–6 to 2005–6

Note: RRBs denote regional rural banks.
Source: For commercial banks from Reserve Bank of India (RBI); for cooperatives and regional rural banks from NABARD.

The share of agriculture in the commercial bank lending experienced an impressive increase, after bank nationalization, from ten per cent in 1975–6 to about 18 per cent at the end of the 1980s (Figure 3.4). However, beginning with the banking sector reforms since the early 1990s, it has steadily declined to reach a low of 11 per cent in the period 2004–6.

Table 3.2
Share of Cooperatives in Total Agricultural Credit in India

(per cent)

Type of loan	1975–6	1983–4	1993–4	2001–2	2002–3	2003–4	2004–5	2005–6
Crop loan	74.9	64.7	69.6	46.4	43.1	41.2	36.7	33.2
Term loan	61.2	40.9	43.6	22.0	16.6	13.2	8.0	6.0
All loans	70.8	56.0	61.3	37.9	34.0	30.9	24.9	21.8

Source: NABARD.

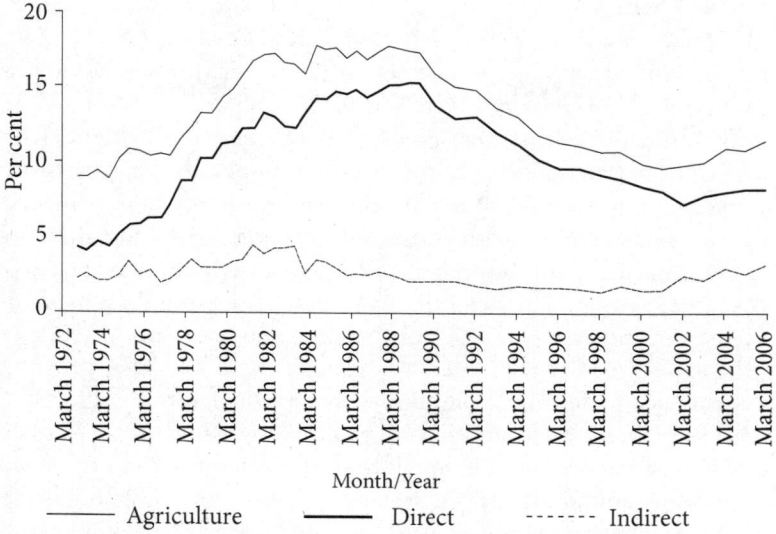

Figure 3.4: Share of Agricultural Credit in Total Scheduled Commercial
Banks' Credit

Source: Computed by *Economic and Political Weekly* Research Foundation based on
RBI, *Basic Statistical Returns of Scheduled Commercial Banks in India*, various years.

Growing Non-institutional Indebtedness of Farmers

The valuable insights provided by the all-India rural credit or debt
and investment surveys historically on estimates of household
indebtedness divided between institutional and non-institutional
sources on a decennial basis are well known. These surveys have shown
a rising share of institutional sources in the indebtedness of the rural
households from about 29 per cent in 1971 to 64 per cent in 1991.
Considering only the cultivator households, the share of institutional
debt had expanded from 31.7 per cent to 66 per cent during the
period. In the 1990s, however, there has been a reversal of the trend
and increase in the share of informal sources in the indebtedness of
rural as well as agricultural households.

   In this regard, the results of three recent surveys on indebtedness of
farmer households have been highly revealing (Situation Assessment
Survey of Farmers (SAS), National Sample Survey Organisation
(NSSO), 2003; All India Debt and Investment Survey (AIDIS), Rural

Finance Access Survey (RFAS), World Bank and National Council of Applied Economic Research (NCAER), see Basu (2005).[3] SAS (2003) shows that of the total of 148 million rural households, 89.35 million (or 60.4 per cent) were farmer households, of whom 43.42 million (48.6 per cent) were reported to be indebted (Table 3.3). As SAS 2003 has covered only the farmer households, its results are roughly comparable with the data on cultivator households provided by the AIDIS surveys. The latter shows a decline in the share of institutional debt outstanding of cultivator households from 66.3 per cent in 1991 to 57.7 per cent in 2003, and a corresponding increase in the dependence of cultivators on moneylenders from 17.5 per cent to 25.7 per cent (Table 3.4). Though there has been growing dependence of farming community on informal sources, there are wide inter-state variations. In Andhra Pradesh, Assam, Bihar, Punjab, and Rajasthan non-institutional sources account for much more than institutional sources (Figure 3.5). What is more revealing is the progressive decline in the proportion of indebted households as well as the share of institutional debt to total debt with the decline in the size of land possessed (Table 3.5). The inverse relation between the

Table 3.3

Number of Rural Households by Farmer/Non-farmer Status and among Farmers by Indebted/Non-indebted Status: 2003

(Number in million)

Rural	Farmer	Indebted farmer	Non-indebted farmer
147.90	89.35 (60.4)	43.42 (29.4)	45.93 (31.1)
		[48.6]	[51.4]

Note: Figures in parentheses are percentages to total rural households; figures in brackets are percentages to total farmer households; farmer household was defined as one in which at least one family member was a farmer, and a farmer was defined as a person who possesses some land and was engaged in agricultural activities on any part of the land during the 365 days preceding the date of survey; and indebtedness refers to liability in cash or kind of Rs 300 or more as value of transaction.

Source: National Sample Survey Organisation, *Situation Assessment Survey of Farmers: Indebtedness of Farmer Households*, NSS 59th Round (January–December 2003), Report No. 498 (59/33/1), New Delhi: Ministry of Statistics and Programme Implementation, Government of India, 2005.

Table 3.4
Relative Share of Borrowing of Cultivator Households from
Different Sources

(Per cent)

Sources of credit	1951	1961	1971	1981	1991	2003
Institutional	7.3	18.7	31.7	63.2	66.3	57.7
Co-op society/banks	3.3	2.6	22.0	29.8	30.0	19.6
Commercial banks	0.9	0.6	2.4	28.8	35.2	35.6
Non-institutional	92.7	81.3	66.3	36.8	30.6	42.3
moneylenders	69.7	49.2	36.1	16.1	17.5	25.7
Unspecified	–	–	–	–	3.1	–
Total	100.0	100.0	100.0	100.0	100.0	100.0

Notes: Borrowing refers to outstanding cash dues, AIDIS, NSSO, 59th Round, 2003.
Source: RBI, *All-India Rural Credit Survey, 1951–2*; RBI, *All India Rural Debt and Investment Survey, 1961–2*; and NSSO, *All-India Debt and Investment Surveys, 1971–2, 1981–2, 1991–2* and *2003*.

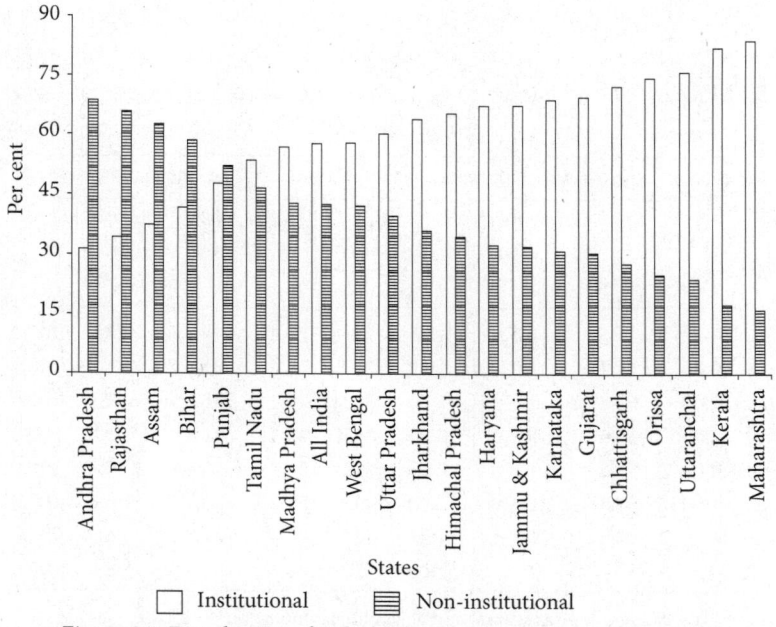

Figure 3.5: Distribution of Debt by Sources across Major States: 2003

Source: NSSO, *Situation Assessment Survey of Farmers, 2003*.

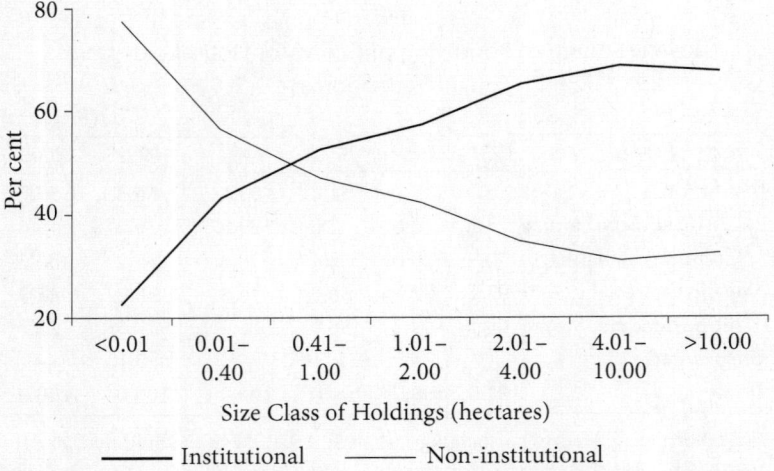

Figure 3.6: Distribution of Debt by Sources across Size Class of Holdings: 2003

Source: NSSO, *Situation Assessment Survey of Farmers, 2003.*

Table 3.5

Distribution of Outstanding Loans by Source across Size Class of Land Possessed: 2003

Size class of land possessed (in hectare)	Institutional agencies				Non-institutional agencies			All
	Total	Government	Co-op society	Bank	Total	Money-lender*	Relatives and friends	
< 0.01	22.6	1.9	5.3	15.4	77.4	47.3	23.1	100.0
0.01–0.40	43.3	4.0	14.5	24.8	56.7	31.8	14.9	100.0
0.40–1.00	52.8	3.8	17.0	32.0	47.2	30.8	9.1	100.0
1.01–2.00	57.6	1.7	20.5	35.4	42.4	25.9	8.8	100.0
2.01–4.00	65.1	1.5	22.6	41.0	34.9	23.4	5.1	100.0
4.01–10.00	68.8	1.3	23.0	44.5	31.2	16.7	5.6	100.0
10.00+	67.6	1.7	23.2	42.7	32.4	17.2	4	100.0
All sizes	57.7	2.5	19.6	35.6	42.3	25.7	8.5	100.0

Note: * Includes both professional and agriculturist moneylenders.
Source: As in Table 3.3.

size of holdings and the share of institutional credit is brought out sharply in Figure 3.6.

The results of RFAS (Basu 2005) are not comparable as the survey covered only two states. However, the results reported provide a telling commentary on the state of access to institutional finance for the vast rural masses. The results are best summed up as follows:

Notwithstanding the progress made over the decades, the majority of the rural population still does not appear to have access to finance from a formal source. According to the RFAS 2003, some 59 per cent of rural households do not have a deposit account and 79 per cent of rural households have no access to credit from a formal source. The problem of access is even more severe for poorer households in rural areas. Indeed, bank branches in rural areas appear to serve primarily the needs of richer borrowers: some 66 per cent of large farmers have a deposit account; 44 per cent have access to credit. Meanwhile, 70 per cent of marginal farmers do not have a bank account and 87 per cent have no access to credit from a formal source. Another segment that faces serious problems in accessing formal finance is the commercial household (i.e. micro-enterprise) segment. (Basu 2005: 4009)

The incidence of interest burden

About 73 per cent of the rural non-institutional debt carried interest rates of more than 20 per cent. About 40 per cent of rural borrowers were paying interest rates of more than 30 per cent on their non-institutional borrowings, while prime lending rates (PLRs) of banks were in the range of 11–12 per cent (Table 3.6).

NEW INITIATIVES FOR EXPANDING CREDIT FLOW TO AGRICULTURE

Concerned at the glaring agrarian crisis, including widespread incidence of farmer suicides and the growing structural constraints faced by the non-farm informal sector, and recognizing the acute shortfall in credit flow to these sectors for over a decade from the beginning of the 1990s, the Government of India, RBI, and NABARD have initiated a number of measures to mitigate the situation. With renewed emphasis on institutional and structural features in credit policy formulation, initially the RBI appointed the R. V. Gupta Committee for studying credit delivery for agriculture (RBI 1998). As brought out earlier, the trends in credit delivery hardly improved for

Table 3.6

Distribution of Debt Outstanding by Rate of Interest for Institutional and Non-institutional Sources as on 30 June 2002, All India

Rate of interest class (%)	Rural			Urban		
	Institutional	Non-institutional	All agencies	Institutional	Non-institutional	All agencies
Nil	1	18	8	3	33	10
less than 6	2	2	2	4	1	3
6–10	4	1	3	12	1	9
10–12	9	1	5	25	1	19
12–15	48	1	28	32	4	25
15–20	34	3	21	22	9	19
20–25	1	33	15	1	18	5
25–30	0	0	0	0	1	0
30 & Above	0	40	17	1	32	8
Total	100	100	100	100	100	100

Source: NSSO, *Household Indebtedness in India as on 30.06.2002, All-India Debt and Investment Survey*, NSS 59th Round (January–December 2003), Report No. 501 (59/18.2/2), New Delhi: Ministry of Statistics and Programme Implementation, Government of India, 2005.

some years up to 2003–4. Consequently, the socio-political pressures were so intense that as many as 18 committees have been constituted to address the questions of bank lendings for agriculture and small and medium-enterprises (SMEs) as well as the institutional issues of the working of regional rural banks (RRBs) and the need for revitalizing the short-term as well as long-term cooperative credit structure. The system of priority sector credit has been looked at afresh; likewise, the norms for regional credit–deposit ratios and investment credit for agriculture have been probed at the instance of NABARD. The RBI's internal working groups have examined the issues relating to rural credit and micro-finance, lendings against warehouse receipts, and the question of adopting 'financial inclusion' as a policy goal. Broadly, these committees and working groups have departed from the traditional methods of targeted lendings and proposed instead a more intensive use of micro-finance institutions along with an innovative system of 'agency banking' as a substitute for branch banking for rural areas. Yet another innovative idea commended by the authorities is that of 'financial inclusion'.[4]

As the recommendations of the various committees and working groups have been by and large inspired by the in-house thoughts on various issues (except the Vaidyanathan Committee [I and II] on cooperatives), they have all been accepted, including those by both the Vaidyanathan committees, and appropriate measures for their implementation have been put in place in official communications. As a result of the acute social pressures to mitigate the travails of the farm community, the Government of India has gone beyond the committee recommendations. To begin with, it has announced a credit package envisaging 30 per cent growth in credit flow to agriculture during 2004–5 and sought to double the total flow over a period of three years from Rs 86,981 crore in 2003–4 to Rs 175,000 crore in 2006–7. As shown in Figure 3.2 earlier, the number of loan accounts in agriculture has risen by 25 per cent from 21.30 million at the end of March 2004 to 26.66 million at the end of March 2005. Besides, they have decided to provide a certain level of subvention to NABARD so that farmers receive short-term credit at 7 per cent rate of interest up to an upper limit of Rs 3 lakh as the principal amount. This measure has sought to accept the principle that interest rate relief to farmers has to be provided through fiscal measures rather than through cross-subsidization within the banking sector (RBI 2006). Also, issuance of Kisan Credit Cards (KCCs), introduced in August 1998, has been yet another step in the direction of expanding farm credit. An average of about 95 lakh KCCs have been issued during the past six years, 2000–1 to 2005–6, taking the aggregate to 591 lakh, of which cooperative banks accounted for the largest share (51 per cent), followed by commercial banks (37 per cent), and RRBs (12 per cent).[5] Hitherto, KCCs were only for crop loans but in 2004–5, their scope was expanded to cover term loans.

The flow of credit to agriculture suffered a setback partly also because credit flow from cooperatives has been sluggish. Obviously, this is what has brought to official focus 'the increasing concern in recent years over the effectiveness, governance and financial health of rural cooperative banks and the attention being paid to rural lending by commercial banks' (Mohan 2006). The implementation of the recommendations of the Vaidyanathan Committee (I and II) designed to revitalize the short-term and long-term cooperative credit structures and minimize the cost of multi-layering, is expected to improve credit delivery.

New Policy Regime for Better Credit Delivery: Intensive Monitoring is the Need of the Hour

The fresh thrust thus conferred on expanding the bank credit base for agriculture and other informal sectors is to be greatly welcomed. However, a close examination of different elements of the new policy regime raises a few misgivings, which are required to be addressed if an enduring impact has to be made on the system of credit delivery for the targeted sectors.

The target for doubling credit flow for agriculture and allied activities in three years and for the SMEs sector in five years appears a knee-jerk reaction to the serious socio-political pressures brought to bear on the system due to vast credit supply gaps created over a prolonged period. During this period, it is not only that the credit flow had dried up; the rural institutional structure in terms of branch-banking had been weakened. Superimposing such a large target on the weak institutional structure will have its repercussions on, first, the quality and purposes of lending and, second, the process of loan recovery. Therefore, attempts should be made to resurrect the entire institutional structure in terms of its geographical spread as well as organizational strengthening. Only such a structure will be able to achieve a steady and healthy delivery of credit for agriculture and rural enterprises.

First and foremost is the need for further spreading of branch network by scheduled commercial banks and RRBs. The system of 'agency banking' could only go to supplement the operations of bank branches in rural and semi-urban areas. A palpable cause for the decline of banks' lendings to agriculture, to small-scale industries, and to small borrowers, has been the banks' professional reluctance towards expanding their branch network in rural areas. As shown earlier, the number of bank branches operating in rural areas (classified uniformly on the basis of the 1991 Census) has experienced an absolute reduction from 33,017 (or 51.7 per cent of the total) in March 1995 to 30,572 (44.5 per cent of the total) in March 2006. Any yardstick we apply—30,600 rural branches serving 5.5 lakh villages, the decline in population per bank office, the period in which bank branches reached break-even points in the past, and the positive externalities they provide—will justify the promotion of rural bank branches. Given the option, the scheduled commercial banks would not like to operate in rural areas. This has been proved clearly since March 1995 after the disbanding of the branch licensing policy and the granting of freedom to bank boards to decide on their branch expansion programme. Since then, there has been a reduction of roughly 2445 rural branches instead of an addition of at least 10,000 bank branches in rural areas under the erstwhile policy thrust. This approach has thus spawned a serious institutional vacuum in the rural credit structure. This has occurred also because no attempt has been made by the authorities to substitute it by strengthening regional rural banks (RRBs) or to build an alternative rural institutional structure for credit delivery.[6] Revitalization of the cooperative sector and RRBs, though recognized as an essential need, will take a decade or more; even thereafter their impact will be meagre compared with the size of resources with the scheduled commercial banks (SCBs).

Second, with vast modern input requirements and diversification into horticultural products and other allied areas underway, agriculture, small-scale industry (SSI), and non-farm activities would require a more sophisticated system of credit delivery for rural and semi-urban branches, for which induction of a sizeable number of qualified agricultural science graduates and graduates with other relevant technical qualifications would be necessary. They may be inducted as a rural banking cadre in RRBs and rural and semi-urban branches of SCBs. Considering this felt need, the renewed policy thrust becomes an excellent opportunity for the government to generate an additional employment of about one lakh posts essentially for rural and semi-urban branches of banks; there are about 3.80 lakh employees in these branches (out of a countrywide bank total of about 8.82 lakh as at the end of March 2004). Of the 3.80 lakh employees, about 1.16 lakh are of officers cadre, and considering the past neglect and the enormous business potential, it would not be too ambitious a goal to induct another lakh of technically qualified officers with moderate salaries befitting rural and semi-urban postings in the next five years or so. A rough calculation suggests that the additional burden of wages and perquisites on this count would work out to about Rs 1200–1500 crore per year after five years; it would constitute less than 0.8 per cent of total income or less than 4 per cent of operating profits of scheduled commercial banks.

Third, it is time that the government bestows on the subject of 'priority sector' a degree of genuineness. Accordingly, a number of large-size loan accounts covered under the 'priority sector' such as housing loans should be eliminated from its coverage. It is necessary that a dispassionate study be made of the items currently covered under the 'priority sector' and the sector be cleaned of items that obviously possess the character of bankability. Fourth, it is necessary to reinforce close coordination between district planning authorities and banking institutions operating in a district. The system of lead bank scheme and associated district-level coordination committees of bankers has apparently become inactive; it needs to be re-invigorated with clear guidelines on respecting the bankers' commercial judgments even as they fulfil their sectoral targets. It is necessary to strengthen the refinancing capabilities of NABARD and to modify the nature of expectations of profitability of rural branches. It is wrong to consider, even as a business proposition, that every rural branch should reach a break-even point and attain positive profits in three years or so. The expectation should rather be to achieve positive profits in a cluster of bank branches, say, within a *taluka* or even a district; the profit so derived should be sufficiently attractive in relation to the totality of business in the whole of the taluka or district.[7]

Further, there is need for systematic monitoring of the effective implementation of various guidelines, both at the level of the RBI and NABARD and also at the individual bank levels. The RBI bestows enormous amount of efforts in monitoring various prudential norms of banks and financial institutions; it is necessary that the RBI assigns the same sanctity to the social goals of banking operations.

Reaching the small borrower also needs mainstreaming the micro-credit system. Today, only women's needs are being catered to and that too to a limited extent through micro-credit. Small borrowal accounts with credit limits of Rs 25,000 or less have accounted for Rs 42,992 crore of loans and the requirements of micro enterprises are estimated to be more than Rs 7000 crore, whereas, out of these small borrowal accounts, the scheduled commercial banks at best may have provided Rs 6900 crore as part of micro-credit arrangement, and that too, with about 80 per cent refinance from NABARD at 6.50 per cent concessional rate of interest. How long can such an arrangement be sustained when the banking system in general shows no commitment to the needs of the small borrowers spread over the nooks and corners of the country? What is being sought to be hypothesized here is that there is a degree of continuum in the economic relationships, say within a village, and the objective of the socio-economic empowerment of the poor households in the village would be better served only if all sections of a village—myriad small and marginal farmers, farm households in general, village artisans, unincorporated enterprises, and other household enterprises—partake the benefits of increased institutional credit but such a requirement is unlikely to be served without co-opting the borrowing needs of all small borrowing households as a responsibility of the banking system and not just micro-enterprises that are self-help groups (SHG) based or are supported by non-governmental organizations (NGOs).

Finally, an important enabling condition for promoting credit delivery amongst banking institutions is the institution of a viable Credit Guarantee Scheme (CGS) in respect of agriculture, micro-enterprises, and other small borrowers.[8] The world over, including in the advanced economies, such schemes are in operation. From the early 1970s, the RBI operated a credit guarantee scheme through its subsidiary Credit Guarantee Corporation of India Limited (CGCIL). The CGCIL operated three guarantee schemes, which covered credit facilities rendered by credit institutions to farmers as well as non-farm informal sector borrowers. After about a decade in 1982, CGCIL also extended guarantee cover for loans advanced by cooperative credit institutions for agriculture and allied activities. By the early 1990s, with claims exceeding fee collections, the RBI had lost faith in the scheme and wrote to the Government of India to discontinue it in April 1990. Though the CGCIL decided to terminate the scheme at the end of March 1991, it was forced to continue the scheme due to some tax considerations on the remaining corpuses. Overall, the credit guarantee arrangement failed because of a socio-economic environment in which loan waiver schemes operated and loan recoveries were difficult. Further, the recommendation of the RBI to discontinue the scheme introduced considerable uncertainty in the minds of credit institutions, and finally, when effective from April 1995, they were given the option to remain with the scheme or to opt out, they soon began withdrawing from the scheme. Now, with the re-emergence of the importance of improved credit delivery for agriculture and non-farm informal sectors, there is scope and essential need for revisiting the credit guarantee system. While comprehensive operational details would have to be worked out by the official agencies, the broad contours of the proposed CGS should cover only loans given by scheduled commercial banks, including RRBs and primary cooperative institutions, for (a) agriculture and allied activities (direct finance), (b) micro enterprises as defined in the new law,[9] (c) artisans, village, and tiny industries, and (d) other informal sector enterprises; it should exclude housing, consumption, and other non-productive loans.

Strengthening NABARD's Refinancing Capabilities[10]

As a refinancing institution, NABARD's resource capabilities deserve to be strengthened. After RBI stopped its contribution to the National Rural Credit (Long Term Operations) (NRC (LTO)) Fund, beginning financial year 1992–3, the only contribution to the Fund has been out of the surpluses of NABARD itself. This has severely constrained the growth of refinance resources available for long-term lending to agriculture.

In case of short-term refinance, the RBI has been providing general lines of credit (GLCs) to NABARD to enable the latter to provide refinance facilities. The GLC limit, which reached a level of Rs 6600 crore in 2000–1, has been gradually tapering off and for the year 2005–6, a GLC of Rs 3000 crore was sanctioned at 6 per cent rate of interest. The RBI's Annual Report for 2005–6 (p. 142) states thus:

NABARD has been permitted to operate the GLC limit sanctioned for 2005–6 up to 31 December 2006. As the limit would not be available after this date, NABARD has been advised to start accessing the markets on a regular basis for sufficient amounts so that the time frame indicated for withdrawal of GLC is adhered to.

Raising resources at market rates and providing refinance would not be a viable proposition for NABARD. There is, therefore, need for continued support to its resources through contribution from RBI to its NRC(LTO) Fund as well as GCL.

The problems of NABARD got further compounded in 2001–2 with the Government of India imposing income tax on NABARD. The ostensible reason was that NABARD is a commercial organization and should be paying income tax to the government. In his budget speech, the then finance minister stated, 'The income of NABARD, National Housing Bank (NHB) and SIDBI was exempted from tax in order to provide fiscal support in the initial years of their functioning. Now these institutions have come of age and are working on commercial lines. I therefore, propose to withdraw the tax exemption available to these institutions' (GoI 2001). To classify a refinancing institution, which has no recourse to cheap retail resources and has to borrow at market rates and extend refinance at sub-market interest rates, as a commercial organization is certainly beyond the realm of simple logic. Imposition of income tax deprived NABARD of the ability to plough back Rs 375–450 crore annually to the agriculture sector.

A refinancing institution cannot depend solely on borrowed funds and has to have access to the resources of the central bank or the treasury. But in the case of NABARD, it has been depending solely on market borrowings since the door had been shut on other funding arrangements. Borrowings, which formed just 11.71 per cent of its loan portfolio at its inception, have now risen to 61.29 per cent. Had Reserve Bank not discontinued its annual contribution to the NRC

(LTO) Fund since 1992–3 and maintained a normal growth level as it had done between 1982–3 and 1991–2 and had the Government not imposed income tax on NABARD since 2001–2, this fund in NABARD would have had an additional Rs 15,000 crore, which would have ensured a higher level of investment credit for agriculture.

Further, in the matter of production credit, if RBI desires to discontinue the GLC, a fund for short-term operations within NABARD has to be created with an initial transfer of funds from RBI and regular contributions thereafter by both RBI and NABARD from their surpluses. The Vyas Committee (2004) stated that in view of RBI's decision to phase out GLC, there would be a need to provide a contingent line of credit at a reasonable rate of interest so that NABARD could draw funds for providing short-term refinance to cooperatives and RRBs. The government, on its part, should restore income tax exemption for NABARD, so that it could plough back larger volumes into agricultural credit.

NOTES

1. This chapter is a follow up of a few detailed contributions made by the author on the promises and performances of the banking industry in India in the recent period (Shetty 2004, 2006a, and 2006b). These studies have provided extensive supply-side as well as demand-side indicators and sought to highlight how practically every indicator has reflected the loss of credit delivery momentum for vast segments of the informal sectors in the post-reform period—agriculture, small-scale industries, micro-enterprises, rural artisans, and other small borrowers. The government has responded to this failure on the part of banks by directing them to double credit for agriculture or SMEs within a short span of three years and by other such measures. The tenability of these measures has also been examined in the studies referred to above.

2. In institutional finance for agricultural development or rural finance per se, one important approach of financing called the supply-leading finance policy perceives finance to play a proactive role. It can be visualized as the creation of financial institutions and extension of their financial assets, liabilities, and related financial services in advance of demand for them, especially the demand from entrepreneurs. Hence, supply leading the demand.

3. First, apart from the usual decennial rural–urban debt and investment survey 2002–3, the National Sample Survey Organisation (NSSO) has

covered the subject of indebtedness also under a special 'Situation Assessment Survey of Farmers' (SAS) conducted during January–December 2003 and published a separate report on 'Indebtedness of Farmer Households' (NSSO Report No. 498). Second, a regular All-India Debt and Investment Survey (AIDIS) has been undertaken for the same period, January–December 2003. Though both of these surveys have covered the same period and have been undertaken in the same NSSO Round (59th), the SAS has defined indebtedness slightly differently; it is 'any liability which was taken in cash or kind is termed a loan, if the amount at the time of transaction was Rs. 300 or more', whereas the AIDIS takes into all cash loans and loans in kind. Finally, there is the 'Rural Finance Access Survey' (RFAS), also of 2003, undertaken by the World Bank and the National Council of Applied Economic Research (NCAER) (Basu 2005). The NSSO surveys are nation-wide surveys, with a major central sample supplemented by a few state/union territory samples, while the RFAS 2003 has covered only two Indian states, namely, Andhra Pradesh and Uttar Pradesh.

4. The findings of the NSS 59th Round (2003) reveal that out of the total number of cultivator households, only 27 per cent receive credit from formal sources and 22 per cent from informal sources. The remaining households, mainly small and marginal farmers, have virtually no access to credit. With a view to bringing more cultivator households within the banking fold, I propose to appoint a Committee on Financial Inclusion. The Committee will be asked to identify the reasons for exclusion, and suggest a plan for designing and delivering credit to every household that seeks credit from lending institutions', *Union Finance Minister's Budget Speech, 2006–7*, pp. 10–11. Accordingly, on 26 June 2006, a committee to prepare a strategy of 'financial inclusion' of the poor was constituted under the Chairmanship of Dr C. Rangarajan. In the meantime, RBI has already put in place a number of measures, including 'no frills' deposit accounts for the poor.

5. It is said that KCCs (Kisan Credit Cards) are not truly cards but borrowal accounts designated as cards.

6. It is reported that through a process of amalgamation, the number of RRBs has been reduced from 196 to 133 as on 31 March 2006 and further to 102 as of January 2007. Now the policy seems to be to rely on RRBs for rural branch expansion; their role in regional spread will remain limited.

7. The suggestions contained in these two paragraphs have also been advanced in another context. See Shetty (2004).

8. There is already a credit guarantee trust fund with Small Industries and Development Bank of India (SIDBI) for small-scale industries (SSIs) but micro enterprises get neglected in this arrangement.

9. The Credit Guarantee Fund Trust for Small-scale Industries (CGTSI) operated by SIDBI has just been revised so as to make the scheme attractive for tiny and rural enterprises. The way it is structured today, SIDBI will not be able to serve micro enterprises and hence the proposed credit guarantee should cover them too.

10. This section relies heavily on a note prepared by P. Satish of NABARD.

REFERENCES

Basu, P. (2005), 'A Financial System for India's Poor', H.T. Parekh finance forum, *Economic and Political Weekly*, Vol. 40, No. 37, 10 September, pp. 4008–12.

Bell, C. and P.L. Rousseau (2001), 'Post-independence India: A Case of Finance-led Industrialization?', *Journal of Development Economics*, Vol. 65, No. 1, pp. 153–75.

Burgess, R. and R. Pande (2003), 'Do Rural Banks Matter?, Evidence from the Indian Social Banking Experiment', CMPO Working Paper Series No. 04/104.

—— (2004), 'Can Rural Banks Reduce Poverty?, Evidence from the Indian Social Banking Experiment', Department of Economics, London School of Economics and Yale University, 9 June.

Burgess, R., R. Pande, and G. Wong (2004), *Banking for the Poor: Evidence from India*, 2004 EEA Meetings, 22 September.

Chadha, G. K. (1993), 'Non-farm Sector in India's Rural Economy: Policy, Performance, and Growth Prospects', V.R.F Series, No. 220, Tokyo: Institute of Developing Economies, October.

Eschenbach, F. (2004), 'Finance and Growth: A Survey of the Theoretical and Empirical Literature', Tinbergen Institute Discussion Paper, TI 2004–039/2, *http://www.tinbergen.nl*.

Government of India (2001), 'Budget 2001–02: Speech of Minister of Finance', New Delhi: Ministry of Finance.

—— (2005), *Report of the Task Force on Revival of Cooperative Credit Institutions* (Chairman: A. Vaidyanathan Committee), New Delhi: Ministry of Finance, Government of India.

Mohan, R., (2004), 'Agricultural Credit in India: Status, Issues and Future Agenda', Lecture delivered at the 17th National Conference of Agricultural Marketing, Indian Society of Agricultural Marketing, Hyderabad, 5 February.

—— (2006), 'Agricultural Credit in India: Status, Issues and Future Agenda', *Economic and Political Weekly*, Vol. 41, No. 11, 18 March, pp. 1013–23.

NSSO (2005), *Situation Assessment Survey of Farmers: Indebtedness of Farmer Households*, NSS 59th Round (January–December 2003), Report No. 498 (59/33/1), New Delhi: Ministry of Statistics and Programme Implementation, Government of India.

Patrick, H. T. (1966), 'Financial Development and Economic Growth in Underdeveloped Countries', *Economic Development and Cultural Change*, Vol. 14, No. 2, January, pp. 174–89.

Ramchandran, V. K. and M. Swaminathan (eds) (2005), *Financial Liberalization and Rural Credit in India*, New Delhi: Tulika Books.

RBI (1998), *Report of the High Level Committee on Agricultural Credit through Commercial Banks* (Chairman: R.V. Gupta), April.

—— (2001), 'Financial Development and Economic Growth in India', *Report on Currency and Finance (1999–2000)*, January.

—— (2004), *Report of the Advisory Committee on Flow of Credit to Agriculture and Related Activities from the Banking System* (Chairman: Vijay Shankar Vyas), June.

—— (2006), *Basic Statistical Returns of Scheduled Commercial Banks in India—March 2005*, Vol. 34, February.

Shete, N. B. (2004), *Role of Lead Bank Officers in the Changed Context of Financial Sector Reforms* (A Research Report), Pune: National Institute of Bank Management, December.

Shetty S. L. (2004), 'Distributional Issues in Bank Credit: Multi-pronged Strategy for Correcting Past Neglect', *Economic and Political Weekly*, Vol. 39, No. 29, 17 July, pp. 3265–9.

—— (2005), 'Regional, Sectoral and Functional Distribution of Bank Credit', in V. K. Ramachandran and M. Swaminathan (eds), *Financial Liberalization and Rural Credit in India*, New Delhi: Tulika Books, pp. 50–109.

—— (2006a), 'Emerging Policy Regime for Bank Credit Delivery and Tasks Ahead: A Critical Review', in R. Radhakrishna (ed.), *India Development Report*, New Delhi: Oxford University Press.

—— (2006b), 'Policy Responses to the Failure of Formal Banking Institutions to Expand Credit Delivery for Agriculture and Non-farm Informal Sectors: The Ground Reality and Tasks Ahead', Paper presented at ICRIER's monthly seminar on *India's Financial Sector*, New Delhi: ICRIER, 14 November.

Singh, S. and V. Sagar (2004), *State of the Indian Farmer: A Millennium Study—Agricultural Credit in India*, Vol. 7, New Delhi: Academic Foundation.

Subba Rao, K. G. K. (2005), 'A Financial System for India's Poor', *Economic and Political Weekly*, Vol. 40, No. 43, 22 October, pp. 4650–1.

Vaidyanathan, A. (1994), 'Employment Situation: Some Emerging Perspectives', *Economic and Political Weekly*, Vol. 29, No. 50, 10 December, pp. 3147–56.

Visaria, P. (1966), 'Structure of the Indian Workforce, 1961–1994', *The Indian Journal of Labour Economics*, Vol. 39, No. 4, October–December, pp. 725–40.

4

Managing Vulnerability of Indian Agriculture
Implications for Research and Development

SURESH PAL

INTRODUCTION

Indian agriculture is in a state of flux. Market-driven diversification in a global perspective has become the new paradigm driving future agricultural growth. Some high-value sectors such as livestock, horticulture, and fisheries are responding to these market forces, while the crop sector continues to stagnate during the last decade or so. There is a consensus that policies and instruments that have served well in the past are either overstretched or are untenable under this new scenario. Therefore, the new growth strategy must be consistent with the important shifts that are taking place. These shifts pertain to price policy, privatization, opening up to international competition, and protection of intellectual property rights. Indian agriculture is responding to these driving forces and showing some signs of stress in the process. Slow growth with rising input costs, deteriorating terms of trade, and dependence of a large population on agriculture are some other major concerns and causes of agricultural distress.

Agricultural distress is a matter of great concern in view of the inexorable downward pressure on farm size and increasing concentration in the lowest size-groups. It is a real threat that if left to market forces alone, poor and small farmers will be left further behind. The declining role of state and deteriorating infrastructure are further causes of concern, especially their impact on the ability to serve smallholders. Equally important are sustainability of natural resources (particularly water) and other environmental externalities, including global warming and climate change. These could cripple our productive capacity sooner than later. Preference for quality products, food and environmental safety, rising capital intensity, and widespread market failure are other forces to put Indian agriculture at a disadvantageous position, and thereby increasing vulnerability of people depending on agriculture. Indian agriculture has to respond to these threats.

It is now believed that the unfolding challenges of Indian agriculture can only be addressed through science and technology, and that adifferent research and development (R&D) paradigm—a national innovation system integrating all facets of rural life and involving stakeholders' participation would be necessary. The agricultural knowledge sector will play a pivotal role in exploiting the new opportunities and containing the likely threats. A new strategy will be required even as we try to infuse vigour and vitality in the existing R&D institutions. There is a need to explicitly incorporate the vulnerability concerns in the R&D strategy and resource allocation. This chapter examines the investment intensity, institutional structure, and orientation of research programmes in the context of reducing vulnerability of Indian agriculture. The chapter is organized as follows. The next section deals with the trends in public investment in agricultural R&D and its regional allocation. This is followed by a discussion on the institutional structure and orientation of research programmes. As a case study of delivery of improved technology, developments in the seed industry and their implications for unfavourable production environments are analysed in the penultimate section. The chapter concludes with a summary of policy implications for making R&D sensitive to the vulnerability concerns.

Public Investments in R&D

Three questions are important for public investment in agricultural R&D: (a) whether public funding is sustained over time, (b) whether the funding is adequate to meet emerging challenges, and (c) whether

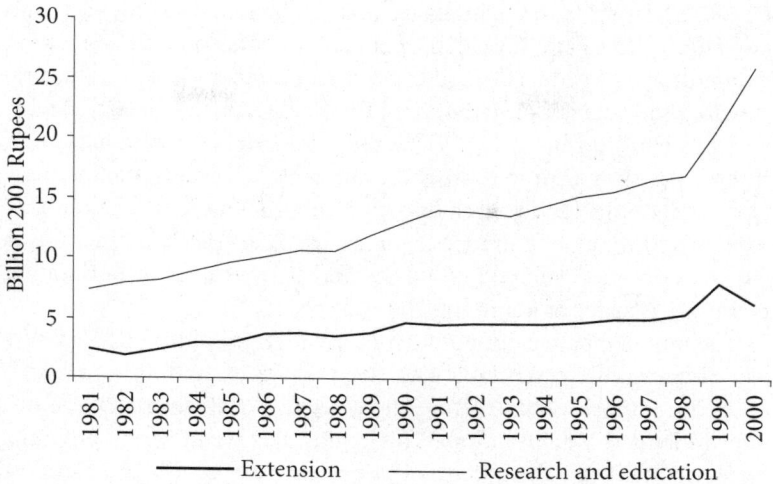

Figure 4.1: Trends in Real Public Funding for Agricultural Research, Extension, and Education

Source: Estimates based on Government of India data.

the allocation is consistent with the development priorities. This section addresses these questions.

Figure 4.1 shows the trends in public funding, in real terms, for agricultural research, extension, and education in India. There is a steady uptrend in the public funding for research and education since 1981, which sharpened further in the late 1990s. This is unlike many other developing countries, except China, which have witnessed stagnation, or even decline in funding. The trend in the funding for extension is rather stagnant. There is, however, a distinct difference in the sources of funding. While most of the funding for extension was from the state governments, they contributed nearly half to the funding for research and education and the rest was contributed by the Central government.

There is no standard measure to assess adequacy of the funding, which is usually derived from R&D agenda. However, a comparison with other countries may be useful, and a standard measure used for such comparison is funding intensity, which measures public expenditure as percentage of agricultural gross domestic product

(AgGDP). In 2000–1, the intensity was 0.42 per cent for research and education (R&E) and it was 0.17 per cent for extension. By separating expenditure on education, the resource intensity on agricultural research in India is estimated as 0.31 per cent, which is nearly double that of extension intensity. While the ratio of research to extension intensity is similar to that observed for other countries, the intensity of funding both for research and extension is much lower than that estimated for other countries. For example, research intensity was 2.6 per cent in developed countries and 0.6 per cent in developing countries (Pardey & Beintema 2001).

The regional pattern of public funding for agricultural R&D shows some interesting trends (Table 4.1). The intensity of funding for research and education is very low for all the eastern states (except Assam) and undivided Uttar Pradesh. This region has tremendous potential for agriculture growth and, therefore, must receive a higher level of

Table 4.1
Intensity of Agricultural Research, Extension,
and Education Funding by States

Region	Research and education (per cent of AgGDP, 2001–2)	Extension (per cent of AgGDP, 2000–1)
Hill region	Himachal Pradesh (1.37)	Himachal Pradesh (0.39)
Northern region	Haryana (0.42), Punjab (0.31), Uttar Pradesh* (0.08)	Haryana (0.16), Punjab (0.02), Uttar Pradesh* (0.24)
Southern region	Andhra Pradesh (0.24), Karnataka (0.42), Kerala (0.52), Tamil Nadu (0.51)	Andhra Pradesh (0.5), Karnataka (0.04), Kerala (0.04), Tamil Nadu (0.6)
Eastern region	Assam (0.39), Bihar* (0.23), Orissa (0.15), West Bengal (0.12)	Assam (0.78), Bihar* (0.23), Orissa (0.15), West Bengal (0.06)
Western region	Gujarat (0.54), Madhya Pradesh* (0.21), Maharashtra (0.59), Rajasthan (0.24)	Gujarat (0.22), Madhya Pradesh* (0.03), Maharashtra (0.26), Rajasthan (0.03)

Note: * Figure is for undivided state.
Source: Pal and Byerlee (2003).

funding. Dryland states with harsh production environments are not necessarily underinvested, and only the states of Madhya Pradesh and Rajasthan have a moderate level of research intensity. The states showing considerable degree of rural distress, for example Andhra Pradesh and Maharashtra, do not have low research intensity either.[1] The same holds true for the extension intensity. However, some of the states, namely Karnataka, Kerala, Madhya Pradesh, Punjab, Rajasthan, and West Bengal have very low extension intensity, even less than half of that for the country as a whole. This deficiency must be corrected.

Although there are interstate variations in the intensity of research and extension, the Indian Council of Agricultural Research (ICAR), which is supported by the Central government, plays an important role in filling the gaps. For example, there is increasing emphasis on eastern India for research, and extension gaps are addressed through the programmes of *Krishi Vigyan Kendras* (KVKs) established throughout the country. It is important, therefore, to recognize the importance of the national level intensity of funding for R&D in influencing the productivity levels in agriculture. Since the benefits of national research spending accrue to different regions, there seems to be no association with the intensity of R&D funding at the state level and agricultural distress in different states. However, one needs to analyse whether research programmes and their outputs were able to address the vulnerability concerns. This is discussed in the next two sections.

R&D Infrastructure

The public agricultural research and education system essentially comprises of ICAR and State Agricultural Universities (SAUs) which together support nearly 22,000 scientists. In 2004, from the aggregate R&D funding, amounting to Rs 28 billion, the ICAR/SAU system claimed more than 85 per cent and the rest was spent by other public and private research organizations. In terms of institutions, the public system has a large network of Central (90) and state (43) research/education units, 126 zonal research stations, and coordinated research programmes (82). The frontline extension system (assessment, refinement, and transfer of technologies) has grown over time, covering demonstrations and KVKs (about 500 in number).

The number of SAUs, that is, the state R&E system continues to grow; these have grown from a mere handful in the early 1970s to 41 in 2007.

Unfortunately, funding levels have not kept pace with this growth in the number of institutions, and operational as well as scientific resources have degenerated. These trends fly against persistent advice and efforts to reorient resources and focus towards local R&D institutions. Establishment of discipline-based SAUs has reduced research integration, thereby further eroding the effectiveness.

Extension or the 'development' component has been under the control of state departments of agriculture. There are no reliable estimates of extension personnel, but it is a known fact that many posts are lying vacant and the staff turnover rate is high, especially in inaccessible and unfavourable areas. Lack of resources, multiplicity of responsibilities, and bias for crop extension are some other notable weaknesses of the system. Consequently, there is tremendous pressure on the frontline extension system of ICAR.

Although the three-tier structure of the research system is supposed to bring it closer to the extension system and farmers, the research–extension–farmer linkages remained rather weak due to lack of an appropriate institutional mechanism. This has in a way resulted into upscaling of frontline extension in the ICAR/SAU system, that is, KVKs and Institute Village Linkage Programme (IVLP).[2] The Extension system, on the other hand, initiated reforms to make the approach more flexible and accountable to stakeholders. Coordination between various line departments and resource generation were also encouraged (ICAR 1998). All these efforts have made some difference, but anticipation of a catastrophic event and preparing farmers to manage it received less priority. This is because the approach to manage such an extreme event needs complete understanding of technology and management portfolio, whereas recent technologies get more importance in technology transfer programmes. What is lacking is a moderate blend of modern scientific knowledge with traditional wisdom to manage not only crop management but also extreme situations such as drought, cyclone, and other weather risks. This becomes more relevant with the growing concern for maintaining sustainability of production systems.

Increasing participation of the private sector, in the post-reform period, has been a significant trend in agricultural R&D. Private R&D is notable in chemicals, including animal health, food, and of late, in plant breeding (Pray & Basant 2001). However, the private sector's exclusive preoccupation with profits restricts its area of interest in a small-farmer dominated agriculture. There is often exaggeration of claims about performance of a technology, and the associated risk, if any, is not made public. This costs dearly to farmers in the years of adversity. Therefore, the presence of a strong and responsible public R&D system is a must, and the tasks of regulation of technology systems and dissemination of information are of critical significance. The persistence of the subsistence farming sector, the flow of a substantial body of new knowledge in the public goods domain, and prevalence of extensive poor and fragile ecosystems are examples to support the need for a strong and effective public R&D. One cannot wish away public–private synergy, but the ICAR and appropriate levels of governments should evolve proactive regulatory systems along with bolstering R&D in agriculture.

Orientation of Research Programmes

Agricultural R&D has considerable time lag in delivering a product and realizing its impact. Therefore, in order to remain effective, an R&D organization will derive its objectives from agricultural development goals and then translate these objectives into commodity, regional, and programme priorities. This process should start well in advance so that timely availability of technology is ensured. Here, performance of agricultural R&D is examined on these parameters, especially targeting of vulnerability concern in research programmes.

The relative emphasis on research, extension, and education has been changing over time. The objective of food self-sufficiency emphasized extension during the initial years of the post-independence era, but during the next three decades, the focus shifted to research and capacity building covering research programmes, institutions, and human resources. Nearly half of the national R&D resources are now spent on research, followed by extension and education. Within research, food crops have been traditional priority commodities. Foodgrains and horticultural crops together account for 40 per cent of the resources. The other major groups such as livestock and fisheries receive nearly 22 per cent of the resources—less than their importance in AgGDP (Jha & Kumar 2006). However, there are significant changes in regional priorities. Rainfed areas, which were rather neglected during the Green Revolution era, got increasing attention in terms of resource allocation and now claimed nearly half of the research resources. However, within rainfed areas, western arid and eastern humid regions were rather underinvested in view of their importance (Pal & Byerlee 2003). Thus, crops and regions vulnerable to various kinds of stresses became important for the system. Sustainability of natural resources, crop management practices to cope with biotic and abiotic stress, and risk management were the main thrusts of the research programmes both for ICAR and SAUs, and, therefore, these received increasing amount of resources from both the systems. As a result, research products for these areas rose significantly. For example, about half of the rice varieties developed during the 1980s and 1990s were developed for unfavourable rice environments (Pal et al. 2005). However, there are no significant breakthroughs in terms of impact on rainfed areas.

The All India Coordinated Research Programmes (AICRPs) have been the major strength of the ICAR/SAU system. These programmes address a research problem in a holistic manner involving relevant disciplines. Integration of research being conducted in different institutions and evaluation of technologies in various agro-climatic zones of the country are other merits of the AICRPs. These progammes have been instrumental in bringing research close to farm realities. A research plan is developed and progress reviewed in annual workshops, which are attended by all the participating institutions, including government departments and the private sector. Technologies developed are evaluated in the coordinated network and performance is reviewed before their transfer. For example, a crop variety is identified for release in the workshop based on its performance in different locations, and the variety is recommended only for those areas where it has performed well. This procedure has impacted the plant breeding priorities—the criteria of tolerance to yield reducers, better grain quality, and wider adaptability are encouraged, besides traditional criteria of yield advantage.

Has the AICRP model been instrumental in incorporating vulnerability concerns in research? This is a difficult question to answer, as there are no reliable counterfactuals. However, one can see that there are AICRPs, such as those for dryland agriculture, water management, problem soils, minor crops, cropping systems, agro-meteorology, etc., specifically dealing with the problems of marginal environments. Programmes of other AICRPs also are organized on an agro-ecological basis and, therefore, all environments are well represented. Results of these programmes have made some impacts and there are studies showing that variability in yield and production of crops has reduced significantly over time. This holds true even for rainfed areas that are prone to various kinds of abiotic and biotic stresses. The spread of modern technology and other associated inputs and infrastructural facilities like irrigation have been instrumental in increasing crop yields with lower variability (Pal & Pandey 2000). This trend, as discussed earlier, is consistent with changing research priorities—there has been higher resource allocation to research for rainfed areas, with a focus on control of yield reducers.

Limitations of R&D Strategy

Though there are efforts to make the R&D system responsive to the needs of unfavourable production environments, there are some notable weaknesses. These pertain to the 'R&D strategy' and response of the system to changing environments. The 'green revolution model' was essentially an input-intensive model, which gradually increased dependence on modern inputs. This has raised the cost of production and capital intensity of agriculture. Farm profitability under this model was sustained as long as crop yields were increasing. But the profitability plummeted sharply with a plateau in crop yields, especially in irrigated areas, causing distress. There were some potential technologies such as biofertilizers and integrated plant nutrient and pest management that are cost-reducing to some extent, but these could not take off because of several reasons. These are information-intensive technologies and, therefore, require special extension efforts and participation of the industry and other development agencies for their dissemination. This is an area where a lot of capacity building efforts are needed.

The input-intensive model was also extended to rainfed areas, primarily practising low input, low output agriculture. Some of the crop varieties tested in the rainfed regions gave better performance and were, therefore, promoted. However, these varieties could not withstand extreme stress (biotic or abiotic) and, therefore, showed larger year-to-year yield fluctuations. These fluctuations were managed, to some extent, with better agronomy in favourable situations, such as assured rainfall. Crops grown in dry rainfed areas with erratic weather and high input use, such as cotton, remained a victim of this model in the absence of varieties tolerant to major pests. Bt cotton could address this problem but high seed cost and lack of information about the technology further worsened the situation. In order to address this problem, the research system, seed industry, and the regulatory system should be in partnership for harvesting the benefits of such technologies. Although the private sector is the main provider of this technology, which has been exploitative in terms of controlling seed prices, policy and regulatory issues assume much more importance to ensure benefits of such technologies to farmers. This is more so when a suitable alternative from the public sector is not seen in the immediate future.

Another major limitation of the R&D strategy is slow progress in terms of flow of new technologies to farmers and their adoption. Part of this could be attributed to appropriateness of the technology, but technology transfer and support systems are equally responsible. Modern technology suffered because of inadequate delivery mechanism and input supply system, and subsistence systems deteriorated due to weakening or breakdown of traditional institutions. As a result, crop yields remained low and increasing pressure on natural resources (land, water, and common lands) degraded and depleted them. The R&D system should work with farmers and other development agents such as civil society organizations for improving the cropping systems. Also, it should work with the industry for tailoring modern technology, such as products of biotechnology, to the needs of farmers. A location-specific approach blending modern science with traditional knowledge could be more appropriate for accelerating the flow of technology and promoting local rural innovations in unfavourable production environments.

Small Farmers

Small farmers have limited resources and poor access to public services such as extension. Their cropping pattern is oriented towards food crops, mainly for home consumption, and livestock is an integral part of the farming system. Therefore, in some respect, their R&D needs, especially extension needs, are different from that of large farmers. The only exception may be biochemical technologies like those seen during the Green Revolution Period, which were neutral to scale, and, therefore, the growth benefits were also shared by small farmers. However, there are other factors that restrict access of small farmers to new technologies. There are increasing capital-intensity and scale bias of technologies and institutional bias in the delivery of technology and associated inputs. For example, mechanical technologies have scale and capital bias and non-availability of credit further restricts access of small farmers to these technologies. Similarly, there are natural resource management technologies which require group action and sometimes long-term investments, limiting the participation of small farmers (Jha 2001). Even the access to biochemical technologies may not be as easy as in the past because of increasing dominance of private sector and rising costs—a trend which is likely to continue in future also. This problem deserves immediate attention. First, there is a need for monitoring developments in technology systems and their implications for the welfare of small farmers. Wherever necessary, the government should intervene to enhance access of small farmers to new technology. Second, the presence of a strong public R&D is essential to make technology systems more competitive and sensitive to the needs of small farmers. This raises a macro policy issue dealing with appropriate roles of public and private R&D in the development and delivery of various kinds of technologies (biochemical, mechanical, crop and resource management, etc.). Trends in private R&D, possible impact of research complementarities and spill-ins, use of domestic and external resources, partnership, etc. should be taken into consideration while addressing this complex issue. The changing roles of public and private sectors in the context of the seed system are discussed in the next section.

Seed System Developments

The Indian seed system has undergone significant changes during the last couple of decades. The most important among these is the increasing participation of the private sector. Initially, private seed companies sold seed of public material and as the industry grew, several large companies diversified into plant breeding. Now the private sector is the dominant supplier of hybrid seed and a significant proportion of the open pollinated variety (OPV) seeds are also supplied by the private sector (Tripp & Pal 2001).[3] One of the impacts of privatization of the seed industry is that there is an increase in the supply of commercial seed, giving a higher seed replacement rate. At the same time, private sector has taken over some of the business of the state seed corporations, resulting in a reduction in their seed volumes. How has this affected price and availability of seeds to farmers, and how could the public system respond under the unfolding scenario in the era of Intellectual Property Rights (IPRs)? This section addresses these questions.

Access to Quality Seed and Information

The performance of the seed industry could be assessed in terms of farmers' increasing access to quality seeds at a reasonable cost and providing them with the required information about seeds. Official statistics show that supply of quality seed of cereals has doubled during the last decade and a half.[4] Also, there is a qualitative change in the sense that increasing proportion of hybrid seed is sold and now hybrids are available for crops like paddy. A recent national survey (NSSO 2005) indicates that nearly half of the farmers purchased seed off-farm (Table 4.2) and the rate of variety replacement was also quite high (Figure 4.2). This trend was witnessed for all the states,

Table 4.2
Use of Commercial Seed

1	Source of seed, all India (NSSO data)		
1.1	Farm saved	52%	
1.2	Purchased	48%	
2.	Paddy, 2002 (NCAP survey)	Andhra Pradesh	Haryana
2.1	Area planted with commercial seed	45%	60%
2.2	Share of private sector in commercial seed	81%	60%
3.	Cotton, 2002 (NCAP survey)	Andhra Pradesh	Maharashtra
3.1	Area planted with commercial seed	59%	90%
3.2	Share of private sector in commercial seed	64%	72%

Note: NSSO denotes National Sample Survey Organisation, NCAP denotes National Centre for Agricultural Economics and Policy Research.
Source: Pal et al. (2005).

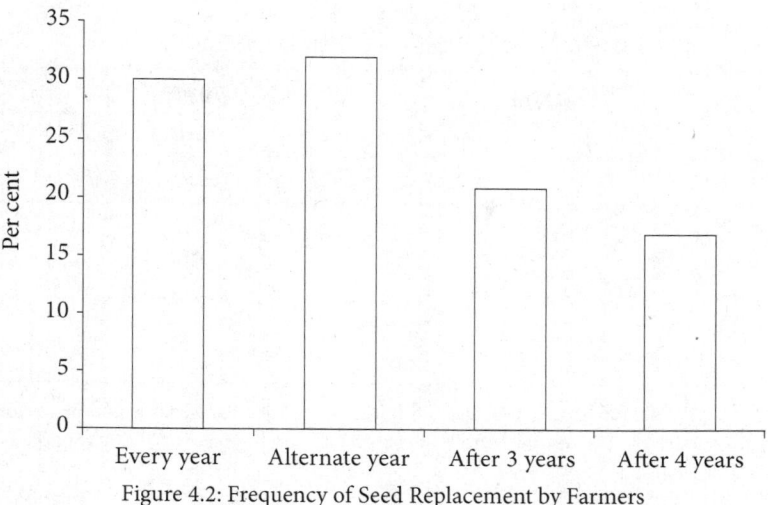

Figure 4.2: Frequency of Seed Replacement by Farmers

Source: NSSO (2005).

except Assam, Chhattisgarh, Jharkhand, and Orissa, where the seed replacement rate was comparatively low—less than half of that for the country as a whole. Our survey conducted in different states also indicates a high seed replacement rate and most of the off-farm seed was procured from commercial sources for quality reasons, even by small farmers. However, the NSSO survey showed a very high variety replacement rate in comparison to our estimates (Pal et al. 2005).

Trends in seed-to-grain price ratio of a crop shows rapid increases in seed prices. There are no reliable time-series data available for this, but the ratio computed using seed cost data of the Commission for Agricultural Costs and Prices (CACP) showed that it increased over time, particularly for the crops sown with hybrid seed. The ratio computed using actual price of hybrid seed for a recent year indicated that it was 8–10 times of that for a variety of the same crop (Table 4.3). Thus, increased availability of seed was associated with an increase in seed prices, which is expected to some extent because of profit margin charged by private seed companies. But this is sometimes too high, often not commensurate with yield or other economic advantage. It is likely that the trend of increasing seed prices will continue in the future also because private seed companies will recover the investment

Table 4.3
Share of Seed Cost and Seed-to-Grain Price Ratio

Crop	Share (%) of seed cost in total cost of cultivation	Seed-to-grain price ratio*			
		Average ratio		Actual ratio, 2003	
		1991–3	1997–2000	OPV	Hybrid
Paddy	6.6	1.34	1.86	3.0	22.6
Maize	7.7	2.11	5.49	3.1	5–13
Cotton	7.6	3.16	6.35	2.2–12.1	37.4**
Gram	26.7	1.53	1.63	2.1	–
Groundnut	30.4	2.43	2.28	1.9	–

Note: OPV denotes open pollinated varieties. * Based on average costs and prices for the major states. For cotton, the ratio is computed using lint prices. ** Ratio for non-Bt cotton seed.
Source: Pal et al. (2007).

on plant breeding and protection of plant varieties. This is an area that the government should monitor closely for making an effective intervention (such as in Andhra Pradesh for Bt cotton) to control any monopolistic tendency.

Although increasing focus on hybrids by private seed companies provides some yield or other economic advantage, there are some costs associated with this trend. There are some good OPVs available in the public sector but these could not find their way in the seed system dominated by hybrids. These OPVs have certain distinct advantages (such as tolerance to stress, and better product quality) over hybrids and farmers may benefit if these seeds are made available. Another noteworthy weakness of the seed system is the lack of information flow to farmers. Private seed companies do not pass on adequate information to farmers about seed and varieties in the market. This is because of two reasons: First, private companies do not have any incentive to promote a public variety which is also being sold by its competitors, and secondly, there is exaggerated claim about expected performance of a proprietary material (Pal & Tripp 2002). Thus, farmers are confused about seed available in the market and often end up by making incorrect choices. The government should play a greater role in the provision of information to farmers as it has the characteristics of a 'public good'.

IPRs and Seed Systems

India has enacted all the legislations to comply with the provisions of trade-related aspects of intellectual property rights of the World Trade Organization. The most important among these for the seed industry is the Protection of Plant Varieties and Farmers Rights (PPV&FR) Act (2001). Amendments of the Patent Act (1970) to cover product and process patents in all fields of technology, especially in the field of biotechnology, also have some implications for the seed industry. The Seed Act is also under revision to harmonize it with these legislations. What will be the likely impact of these legislations on the Indian seed industry?

The impact of the new IPRs regime can be realized only when there is adequate institutional capacity to implement and enforce it. Such capacity is almost in place as far as administration is concerned, but development of capacity within public plant breeding programmes and enforcement agency requires some more efforts. Assuming that the enforcement will be in line with other property rights and the overall working of the judiciary system, some expected impacts of IPRs on the industry could be discussed. The impact on plant breeding priorities, relations of public research with private seed companies, and transfer of technologies by multinational companies (MNCs) deserve special attention. It is believed that there may not be further diversification of plant breeding priorities in the private sector, as the PPV&FR Act does not provide any further incentive for this. Private seed companies fear that farmers, under the provision of saving, exchange, and sale of crop produce, can directly compete with them, thereby reducing their incentives. Therefore, private plant breeding may be confined to hybrids only. However, there could be a change in public–private relations if the public programmes visualize IPRs as an opportunity to generate resources and opt for some kind of relationship to meet this objective. Small players may not find it easy to survive in the market without the support of public plant breeding. Therefore, public research should use the IPRs regime to foster partnership with the private sector, accelerate transfer of technology, and promote competitiveness of the seed industry. In this process, there is a possibility that the public programmes can generate some resources. Therefore, the public sector should learn to manage its IPRs to strike a balance between the efficiency (serving commercial farmers) and equity (meeting the needs of small farmers in unfavourable areas) objectives.

Assessing the impact in terms of technology transfer by MNCs and concentration in the industry is rather difficult at this stage because this will be largely influenced by the degree of effectiveness of the IPRs regime and it is too early to make some assessment on this. However, there could be a transfer of some technologies from overseas to increase market share and these technologies could be protected by some combination of IPR and non-IPR mechanisms such as that followed for Bt cotton. This may result into some sort of concentration in the industry—MNCs owning cutting-edge technology will dominate the market and determine seed prices. The government should monitor the industry for such a trend and intervene effectively in case monopolistic power is likely to emerge. Finally, the impact of IPRs should be seen in conjunction with economic policies and other regulations such as seed and bio-safety rules, which are also important for the growth and diversification of the industry. These factors were primarily responsible for rapid growth and diversification of the Indian seed industry in the non-IPR era (Pal et al. 2007).

Delivering Seed of 'Orphan Crops'

With the decline of the public seed system, delivery of OPV seeds of the crops with high seed rate, often called 'orphan crops', has became a challenge. The case studies for groundnut and potato have shown that the variety replacement rate is very low and farmers get little information about new varieties from public extension or the seed system. Although the seed replacement rate is quite high for these crops, the formal seed system is able to meet hardly one-third of the seed requirement and the rest is met through farmer-to-farmer exchange of seed. Concerted efforts are required to augment seed supply, improve seed quality, and promote new varieties. Technological innovations should address major production constraints such as aflatoxin in groundnut and improve the multiplication rate and quality of seed. It is very unlikely that these crops will attract private investment in plant breeding in the near future even under the new IPRs regime because seed saving is quite common for these crops. Therefore, the public research system should continue to shoulder the responsibility of plant breeding, and develop partnership with the private sector to promote decentralized seed activities. The state seed corporations should also strengthen their seed business with focus on better seed multiplication and quality management (Pal et al. 2006). Coordination among public seed agencies may provide opportunities for cost reductions, augmenting supply in deficit regions, and offering greater choice to farmers. Initiatives such as mechanisms for sharing resources, improving private sector access to public research material, encouraging public–private joint programmes, and involving the private sector's participation in policy-making will foster the partnerships. Since both the sectors have incentives (public sector to increase the research impact and the private sector to commercialize public varieties and reduce cost), such a partnership is likely to be sustainable.

Road Map

This chapter has examined the trends in resource allocation and orientation and strategy of agricultural R&D. It is shown that unfavourable production environments such as dryland areas and management of vulnerability of Indian agriculture have received increasing attention in resource allocation and targeting of research programmes. Rainfed areas now account for nearly half of the national resources and the states witnessing stress among farming community are not underinvested. Management of extreme events such as drought requires pooling of all knowledge and resources and should primarily be the responsibility of the extension and other state line departments, which need to develop such capacity in terms of more resources and trained manpower. Of course, the research system will be an important ally in this task in the provision of required technology and expertise. Sometimes modern technology needs to be used in combination with indigenous knowledge and, therefore, the research system will be helpful in educating the extension system and farming community in this regard. Increasing profitability and farm income by cost reduction or yield improvement and promoting sustainable and equitable use of natural resources are other important but long-term research issues. Any contribution to these development objectives will help absorb shocks that are likely to reduce farm income drastically. These issues, therefore, should continue to receive high priority.

Restricted flow of information about technologies and market trends is a serious limitation. The government must address this problem. This requires building capacity to collate information, anticipate trends and shocks, and disseminate enriched information to exploit growth opportunities and neutralize the adverse impacts. Here it may be noted that besides traditional yield risk, new sources of risks are emerging and important among these are the risks associated with adoption of new technology, erratic product prices, non-availability of essential inputs, etc. Management of these new risks should be an integral part of the development approach.

Economic viability of small farmers is a major concern, and all measures should be used to address this. Agricultural R&D should aim at addressing the needs of small farmers. Technologies for reducing the cost of cultivation, providing an alternative to those crops and products that are likely to lose under the new trade regime, options to participate in high-value agriculture, and the development of value chains deserve special attention. Capacity building for addressing these and other emerging issues is accorded high priority under the Indo-US Agriculture Knowledge Initiative and other externally-funded projects. It is likely that capital-intensity of technology will further increase, both because of capital-biased technological change and increasing participation of the private sector in R&D. In order to facilitate access of small farmers to such technologies, delivery of institutional credit should be strengthened. The regulatory system for technologies such as seeds of improved variety will also have a significant impact on farmers' access to such technologies. Therefore, the regulatory mechanism should be credible and effective. Finally, any measure to promote agricultural development and manage vulnerability will be successful when it is supported with efforts to reduce the dependence of population on agriculture.

NOTES

1. For a detailed discussion on sources, trends, and regional pattern of funding, see Pal and Byerlee (2003).

2. As slight variation in conditions could turn technology irrelevant, IVLP links the ICAR institute's front-line extension to villages for technology assessment and refinement.

3. Seed of an open pollinated variety can be saved for sowing in the next season, while farmers need to buy seed of hybrid in every season, thus attracting greater private sector participation. Both the terms are used

together to indicate this difference, otherwise use of variety alone as a generic term indicates both types of seed.

4. Quality seed refers to commercial seed, including both certified and truthfully labeled seed.

REFERENCES

ICAR (1998), 'National Agricultural Technology Product: Main Document', New Delhi: ICAR.

Jha, D. (2001), 'Agricultural Research and Small Farms', *Indian Journal of Agricultural Economics*, Vol. 56, No. 1, pp. 1-23.

Jha, D. and S. Kumar (2006), 'Research Resource Allocation in Indian Agriculture', Policy Paper 23, National Centre for Agricultural Economics and Policy Research (NCAP), New Delhi.

NSSO (2005), *Situation Assessment Survey of Farmers: Some Aspects of Farming*, NSS 59th Round (January–December 2003), Report No. 496 (59/33/3), New Delhi: Ministry of Statistics and Programme Implementation, Government of India, July.

Pal, S. and D. Byerlee (2003), 'The Funding and Organization of Agricultural Research in India: Evolution and Emerging Policy Issues', Policy Paper 16, New Delhi: NCAP.

Pal, S. and S. Pandey (2000), 'The Nature and Causes of Changes in Variability of Rice Production in Eastern India: A District-level Analysis', in S. Pandey, B. C. Barah, R. A. Villano, and S. Pal (eds), *Risk Analysis and Management in Rainfed Rice Systems*, Limited Proceedings No. 5, International Rice Research Institute, Philippines.

Pal, S., H. Singh, and P. Mathur (2005), 'Indian Seed System Development: Policy and Institutional Options', mimeo, New Delhi: NCAP.

—— (2006), 'Delivering Seeds of "Orphan" Crops: The Case Studies of Potato and Groundnut in India', Paper contributed to 26th Conference of the International Association of Agricultural Economists, Queensland, Australia, 12–18 August .

Pal S. and R. Tripp (2002), 'India's Seed Industry Reforms: Prospects and Issues', *Indian Journal of Agricultural Economics*, Vol. 57, No. 3, pp. 443–58.

Pal, S., R. Tripp, and N. Louwaars (2007), 'Intellectual Property Rights in Plant Breeding and Biotechnology: Assessing Impact on the Indian Seed Industry', *Economic and Political Weekly*, Vol. 42, No. 3, 20 January, pp. 231–40.

Pardey, P. G. and N. M. Beintema (2001), *Slow Magic: Agricultural Research a Century after Mendel*, Washington: IFPRI.

Pray, C. E. and R. Basant (2001), 'India', in C. E. Pray and K. Fuglie (eds), *Private Investment in Agricultural Research and International Technology Transfer in Asia*, AER-805, Economics Research Service, Washington: USDA, *http://www.ers.usda.gov/publications/aer805*.

Tripp, R. and S. Pal (2001), 'The Private Delivery of Public Crop Varieties: Rice in Andhra Pradesh', *World Development,* Vol. 29, No. 1, pp. 103–17.

Farmers' Distress:
A Few States in Focus

5

Farmers' Distress in a Modernizing Agriculture—The Tragedy of the Upwardly Mobile

An Overview

THE EMERGING AGRARIAN SCENARIO

Prime Minister Manmohan Singh sounded a sombre warning on agriculture in the meeting of the National Development Council held in December 2006 to consider the strategy and priorities for the Eleventh Five Year Plan of the country. He pointed out the fall in the growth rate of agriculture since the early 1990s and emphasized the importance of achieving a growth rate of 4 per cent per annum in agriculture during the Eleventh Five Year Plan to stabilize the Indian economy on a high growth trajectory. At the same time, activists such as Medha Patkar of Narmada Bachao Andolan and the Naxalite movement carry on relentless struggles to protect the livelihoods of the millions in agriculture who toil for agricultural growth and, in return, get pushed to the brink of subsistence. While the Prime Minister focuses on crisis in agriculture, the activists are driven by the agrarian crisis reflected in growing landlessness and casualization of labour in agriculture, unchecked proliferation of small and marginal holdings, and widening rural–urban gap in development. Agrarian crisis is eroding the economic and social foundation of rural India and threatening the stability of the Indian economy and polity. These two views—crisis in agricultural growth and agrarian crisis—are overlapping perspectives but they lead to sharply differing diagnoses of the problems in agriculture. While the former favours liberalization, globalization, and a less interventionist role of the State, the latter firmly believes in the need for radical changes in the structure of society through land reforms and political mobilization of the poor.

Recent indications are that India is unlikely to veer towards either of these two extreme ideological positions. It is far more probable that the country works out a middle path with the two ideologies accommodating each other. The Eleventh Five-Year Plan is likely to witness a major step forward in public and private investment in agriculture and in improvements in critical services such as credit, extension, and insurance. At the same time, the Central and state governments would be under considerable pressure from activists, proactive judiciary, and political mobilization of the poor to take concrete steps to implement the constitutional scheme for decentralization, which has received only lip service so far. The role provided for Panchayati Raj Institutions (PRIs) in the National Rural Employment Guarantee (NREG) programme could be a well-planned first step in that direction. Meanwhile, self-help groups (SHGs) have made encouraging progress in bringing together the rural women from poor households for collective action in the fields of saving, thrift, and income generating activities. If the hardcore poor become active and learn to demand their dues in benefits from development, there can be several positive impacts—more effective safety nets, growing participation of the poor in local governance and development planning, and building-up of pressures from below on the higher-level governments for more power and resources. A recent example is the comprehensive social security scheme which at the moment (July 2008) is only on paper but commits the government to implement measures going far beyond the safety nets implemented so far. All this is still largely in the realm of hope but it has more substantial base than mere wishful thinking.

A worrisome aspect of the prospects in agriculture is that there is still no assurance about the rise in the growth rate or about the growth being sufficiently inclusive in nature. Politicization of the poor without a matching response from growth in agriculture could bring in numerous interruptions and breaks in the development processes, creating the vicious circle of uncertainty generating pessimistic expectations, which, in turn, result in more uncertainty. Growth in agriculture now depends on millions of producers with modest holdings and resources and located in a milieu far from congenial to growth. Modernization of agriculture is still limited to a few areas that are relatively better-off in this respect. It is in these very areas providing a role model for growth that farmer distress has emerged and spread rapidly since the early 1990s in the wake of reforms for liberalization and globalization. Farmers' suicides in these areas have highlighted the distress and have spread panic widely, affecting even the central and state governments. Farmers' suicides indeed need prompt relief and rehabilitation programmes. A measure of panic on this count could even be a positive factor serving to speed up these programmes. However, farmers' distress, which is the huge iceberg hiding below the visible tip of suicides, could be a far more serious growth-retarding factor in the long run than the suicides themselves. It is hardly a coincidence that a recent round of the national sample survey (59th Round, January–December 2003, National Sample Survey Organisation [NSSO] 2005) brought to light the disturbing statistic that a very high proportion of the farmers in the country would prefer, if given a chance, to leave agriculture! Suicides make headlines for a while yielding ground to other more newsworthy events. The public memory is short and the capacity of the governments to forget is legendary. It is, hence, important that one takes a careful look at the prevailing farmers' distress, traces its roots, and identifies the broader societal context generating the farmers' distress.

Three-tiered Pyramid

It is useful at this point to view the farming communities in India as forming a three-tiered pyramid. At the wide base are located the poverty-struck farmers who hover around the poverty line and remain in distress irrespective of the state of weather and market. The middle tier is formed by the upwardly mobile farmers engaged in a prolonged and, often, frustrating battle to rise above the poverty line and join the mainstream. Small but politically powerful farm lobbies occupy the peak of the pyramid. It is important to note that this pyramid is an outcome of policy regimes that have prevailed so far in the Indian economy and the rural and agricultural sectors. A striking feature of the pyramid is that none of the tiers have strong orientation towards growth and development. The base is engaged in a precarious struggle for subsistence. The middle tier operates under numerous constraints and is compelled to choose risky ventures with potential of quick gains, which often turn out to be illusory. Many among them sink rather than swim. As regards the top, the farm lobbies find it more profitable to look for subsidies and free lunches rather than make use of their impressive entrepreneurial potential and risk-bearing capacity for agricultural growth and modernization. A sad feature on Indian society and polity is that the relatively better-off mainstream in the society has far more lobbies than honest and hard-working entrepreneurs. An even sadder feature is that the more elitist the lobby, the more dysfunctional it tends to be in its economic and political role! Lobbies are legitimate entities in a democratic polity. Farm lobbies are bound to grow in number and size in the coming years. As those in the middle tier and at the base get politically mobilized, new relationships and balances between the existing lobbies at the top and the newly forming lobbies would emerge. It is this process which will curb the dysfunctional features of the present lobbies and induce all the lobbies to realize the importance of self-discipline and respect for economic constraints on subsidies, rents, and free lunches. The experience so far shows that the government is quite powerless to regulate the lobbies, particularly those at the top, which, in fact, dominate the government decisions and policies. It is only when the lobbies expand to include the non-elites that the government would become truly democratic and genuinely committed to equity-promoting goals such as poverty elimination and supporting the weak and the disadvantaged.

THE UPWARDLY MOBILE FARMERS

Keeping this societal context in mind, the approach adopted in this book is to focus on the upwardly mobile farmers in the middle tier and analyse the policies needed to support them. This is a strategic rural stratum in the sense that if it gets activated and contributes to growth, there is a good chance that growth could become inclusive, expanding the economic space available for those at the base of the pyramid. It needs to be remembered that the formidable barrier at present in poverty elimination is the meagre economic space available for the poor. If this space expands steadily and substantially, safety nets for the poor, PRIs, SHGs, and other institutions empowering the poor, and ability of the poor to participate in the mainstream markets, institutions, and networks would all improve. The process of poverty elimination would gather thrust and momentum. Farmers' suicides have been large in numbers and spread widely across areas outside Green Revolution agriculture. If these farmers become active and viable, the resulting agricultural growth and its linkages would benefit areas and farmers who have been bypassed so far by growth and development processes. These upwardly mobile farmers would also play a role in diminishing the present influence of farm lobbies and lobbies of urban elites on the government strategies for growth and development. Since the numbers of these farmers would be too large to be co-opted through subsidies and free lunches, the government support to them would have to be much more genuinely growth-oriented than in the past. Further, as the upwardly mobile farmers themselves emerge as a lobby, the political clout of the existing lobbies will tend to get eroded.

Part II of this volume contains five studies of farmers' distress and suicides in five selected states (Andhra Pradesh, Karnataka, Kerala, Punjab, and Maharashtra). Besides a look at the state as a whole, these studies cover areas in each state undergoing rapid modernization of agriculture of the type which the Indian Five Year Plans prescribe for raising the growth rate in agriculture. Farmers' suicides in these areas have been too numerous and frequent to remain hidden in the general statistics on suicides. As may be expected, suicides usually have many precipitating factors. But, the state studies included in this part clearly bring out the role played by farmers' distress in the case of suicides by farmers. While the situations in these states have their own specificities, a commonality among them is the growing pressure of indebtedness, rising costs, and declining returns. Another commonality is the inadequate policy support to farmers precisely when their need for support was most pressing. Even institutions lending finance remained indifferent to the farmers' woes. One would legitimately expect these institutions to remain alert to the financial conditions of their borrowers and to intervene before the borrower reaches the point of crisis. It is these two factors—inadequate policy support and unsympathetic and unhelpful institutional lenders—that need a serious look while preparing a road map for the future.

Both nature and markets are too capricious to be controlled fully by the policy maker. The shocks they generate are random in their occurrence and are powerful enough to shatter the household economies of millions of farmers with meagre capacity to survive the shocks, particularly when the shocks bunch together over time. This is the context that inhibits growth in agriculture and, more particularly, in areas with modernizing agriculture.

PROFILES OF FARMERS IN DISTRESS

Out of the five states mentioned above, farmer suicides have been particularly numerous in some districts in Andhra Pradesh, Karnataka, Kerala, and Maharashtra. The Prime Minister visited some of these areas and announced a special package of relief in July/August 2006 in 31 districts of these four states. Despite this, suicides have not been halted. The following cases from the state studies included here provide glimpses of the farmer households experiencing severe distress.

Box 5.1: Case Studies of Farmer Households with Incidence of Suicides in Maharashtra

Case A: Forty-five year old male. The household has 6 acres of land and has an annual income of Rs 35,000. Discussion revealed that: (i) delayed monsoon led to double/triple sowing, increasing input costs for the household; (ii) the deceased individual had entered into cotton trading, for which he had taken loans and invested on other farmers, and as a result the impact of crop failure by other farmers would also add to his burden; (iii) the individual had plans of getting at least one of his daughters married; (iv) he was also contemplating contesting the local Sarpanch elections, indicating that any economic downfall would affect his social reputation immensely. From this case it can be inferred that there is a complex interplay of multiple causes that are not mutually exclusive. After the demise of this individual there were three/four cases in nearby villages in about 7–10 days. This suggests a cascading effect.

Case B: A young man in his early 20s. A few years ago his father was not well and he took over the reigns of cultivation and experimented with input intensive cultivation. From the returns he could spend on his father's treatment, expenses on a sister's marriage, and improve

(Contd.)

(Box 5.1 contd.)

the overall economic condition of the household. Other farmers also started taking advice from him. He was confident of a good crop and initiated plans for getting the other sister married. He had also told his mother that he would like to get married to a girl from a poorer household without taking any dowry. To his dismay, the crop failed ...

Case C: Fifty-two year, male. They had 52 acres of land at one time, but now they have 36 acres only—indicating a scenario where the economic position is declining. The individual had outstanding loan from formal as well as informal sources and a couple of days before the fateful incident, the private moneylender had visited and insisted on being returned the money or transferring ownership of some land. There were two daughters of marriageable age. There was a burden on higher education of his children (daughters were pursuing their graduation and post-graduation, a rare thing in this part of rural Maharashtra, and sons were doing their computer and management courses). The deceased had been talking some time ago with his children about other suicide cases and has given his opinion that it was not a solution; this perhaps gives a faint indication that he might have been contemplating suicide for some time.

Source: Mishra (2006).

Box 5.2: Cases of Farmer Households with Incidence of Suicides in Andhra Pradesh

Case A: Shankar was ambitious, and wanted to live a good life. When we were in a joint family the main occupation was toddy tapping and cultivation. We got separated; we also purchased a share of the toddy trees (5–6 trees) for Rs 3,000. The income was just sufficient for our sustenance. As the family expanded, the income was not sufficient and Shankar took two acres of land on lease for 2–3 years. In the first year, he planted cotton in one acre and then extended it to two acres. He also planted chilli in two acres in one year. We had no bullocks or plough and worked with a rented tractor. It went on smoothly for a couple of years. Later he purchased half acre of land and then another quarter acre for which he borrowed Rs 30,000 from private sources. After purchase of land he went for bore well which yielded hardly any water. He also went for an open well around the bore well to a depth of 30 feet, which cost Rs

(Contd.)

(Box 5.2 contd.)

17,000. He purchased a motor for Rs 3,000. All this he did within a span of one year. He was one who never shirked work and was always thinking about how to consolidate the agriculture. All this led to a cumulative debt of Rs 1.1 lakh, which became burdensome. One of our relations pressed hard for the money and insulted him in the public. Unable to bear the pressure he consumed pesticide in the house. (Manda Rama Devi, wife of Manda Shankar from Chityal village in Warangal district)

Case B: Balakondanna's father had four acres of land. He and his two brothers got one acre and thirty cents each. In addition to that he purchased 4 acres of land in 1997. They were cultivating groundnut, kusuma (oilseeds), and paddy. Paddy was cultivated in the lands under the canal irrigation. They have not invested on bore wells. The crops failed for four consecutive years. The loans increased to Rs 2 lakh due to high interest rate. He sold 100 sheep to clear the loan of Rs 1.0 lakh. There was poor crop in the preceding two years. He borrowed loan on 24 per cent annual interest. He could not clear the debts and the interest and principal became too much. There was still a lakh rupee loan remaining. The pressure was mounting and those who gave the loan were frequently coming home demanding the repayment. There was great force from the moneylenders. 'To make a living we had to cultivate for which we had to borrow. But our hopes were crushed, leaving only worries. The man is gone.' (Bandi Meenakshama, 36 years, Khurba caste [Backward Communities—BC] from Bukkarayasamudram village Anantpur District.)

Case C: 'My husband had not inherited any land from his father; he was taking land on lease from a trader in the village to an extent of 5 acres. He had debts from the beginning by the time of our marriage itself. We both worked hard to repay the loans, which were taken for my step daughter's wedding, ceremonies, etc. Six to seven years back (around 1997) we purchased land from the trader whose land we were cultivating on lease, for 1 lakh rupee. We constantly saved money to buy land, and with the help of a loan, we were able to invest one lakh rupees in purchasing 5 acres of land. This loan was taken from the Primary Agricultural Co-operative Society (PACS) (long term loan). We cultivated cotton, chilli, maize, etc. in our fields. For two years the yield was good. We had invested a large amount of money in digging borewells. At a time we went for five bore wells (around 1998), three

(Contd.)

(Box 5.2 contd.)

failed and two are working. We thought we could repay the debt quite easily as we would get good water which would help to increase the produce from our field. The debt is increasing day by day and we did not know what to do. The crop was never satisfactory in those years. We also had to meet other expenses for our daughter. We took a loan from the bank (PACS), to the extent of Rs 1.5 lakh for digging of the bore wells. The bank people were pressurizing every day. They gave us several notices, which my husband could not stand. He went one day in the morning, sold out the land for 1 lakh rupees and repaid the bank amount (1998).' Seetha Reddy managed to clear the bank loans by selling their land but could not clear the loans taken from the moneylenders. They abused us saying that we paid the bank dues but not their dues. (Bathula Ramanamma aged 44 years, Reddy caste, Gangapur village, Jadcherla Mandal, Mahabubnagar district.)

Case D: Devarapalli Satyanarayana is a small farmer having 1.5 acres of land located in Deepaladinnepalem, in Sattenapalli Mandal of Guntur district. He belongs to the Kamma (forward community) caste. In addition, he took 4.25 acres on lease. All the area operated by him has assured irrigation (canal irrigation). The fixed rent on the land is Rs 6,000 per annum per acre. He cultivates paddy in 2.75 acres and chilli in the rest 3 acres. The predominant problem he faces is insufficient water because of the location in the tail end of the canal system. Some time back he also cultivated chilli and maize. Because of insufficient water he restricted the paddy crop. 'Also because of non remunerative prices I switched over to sesame and maize from cotton' said Satyanarayana. He has a debt of Rs 72,000, of which he could repay Rs 17,000. Because of low price he had not yet sold the produce at the time of interview. The major portion of his debt is from regional rural bank and government departments. Only around 12 per cent of the debt is taken from non-institutional sources at an interest rate of 24 per cent per annum. He says that the Village Agricultural Officer (VAO) extends extension services and sometimes he also depends on the fertilizer dealer for advice. The major problems faced according to him are insufficient extension for pest attack, irrigation, and price crash. Crop diversification, less dependence on non-institutional source of credit, and having milch animals are some of the positive aspects not pushing him to the brink of distress.

Source: Revathi (2007).

Box 5.3: Case Studies of Distressed Farmer Households in Kerala

Case A: The story of Ramakrishnan, a forty-five year old farmer, is a classic case of agricultural indebtedness. His family includes his wife and two sons, one son working in a milk society on daily wage basis and the other, a B.Com. student. They have a land holding of 1 acre and 50 cents dryland and 35 cents of wetland. In the wetland, the family cultivates rice, the entire production going for their own consumption, which lasts for six months. For the rest of the year, they have to purchase rice. According to Ramakrishnan, in earlier years, the income from agriculture had been sufficient for the family. One portion of his landholding was under rubber cultivation. In 1996, seeing that rubber prices were steadily going down, he had begun to use the rubber trees as a support for pepper vines that he decided to expand. It had appeared to be a wise decision and he was proved right. Within three or four years, the income from pepper production increased and he was able to repay his debt. The price for pepper was then at a high of around Rs 150 to Rs 175 per kilogram. Ramakrishnan's rough estimation was that he could get an yearly income of Rs 40,000 from pepper and save some amount as well. Hoping that the good price for pepper will remain for a long period, he constructed a house by taking two loans, one from the State Bank of Travancore for Rs 150,000 and another from Pulpalli Service Co-operative Bank for Rs 25,000. But the favourable agricultural situation did not last long. The quick wilt and the slow wilt diseases ruined his pepper vines and his financial situation deteriorated and he was not able to repay the bank loan on time. Adding to the worries of the family, Ramakrishnan suffered a heart attack and was hospitalized. Fortunately for them, he got free treatment from Sree Satya Sai Baba hospital and did not have to spend much on that account except small amounts on travel and expenses for the attendant's stay in the hospital. Two years earlier, he had planted 250 Areca nut plants on his land. But during the severe drought of 2004–5, he lost all the areca nut plants and also other crops. Though he got some support through the drought relief scheme (Rs 6000), it was simply not sufficient. In 2004, he cut down and sold some silver oak trees and repaid Rs 40,000, which he owed to the bank. Now he does not have any option. He is now looking after the shop of the Self Help Group in which he is a member.

(Contd.)

(Box 5.3 contd.)

Case B: Sukumaran, 45 years, is a farmer with three acres of own land. He has a wife and two daughters. His main cultivation is pepper. In 1995, at a time when the agricultural situation was good and fetched sufficient income to the family, he borrowed Rs 50,000 from his friends on pay-back condition to cultivate ginger at Coorg. With this money and his income from pepper, Sukumaran started ginger cultivation in 3 acres of land that he leased-in in Coorg. Later he extended the ginger cultivation to 12 acres of land. By the year 2000, the quick wilt started attacking pepper vines. As he was seriously involved in the ginger cultivation over such a vast area, he could not provide proper care to his pepper vines in his own lands. This accelerated the crop loss and loss of income. Income from pepper had been a sure source for him and it had helped him invest in ginger cultivation on leased land. When his endeavour failed, Sukumaran could not pay back the borrowed money. In 2001, he took Rs 100,000 from the Indian Bank, Pulpalli, pledging his land, half of which he used to pay back the amount borrowed from his friends and the remaining half for settling his various other liabilities contracted for taking care of a variety of needs of the family. With the failure of pepper cultivation due to disease, and price fall of ginger, he defaulted payment of the bank loan which kept accumulating and increased his liability.

Case C: Joseph, 66 years, owns three acres of land and cultivates rice, ginger, pepper, and coffee. Cultivation is his only source of income. After 2000, he observed production from his crops drastically decreasing due to crop diseases. Price fall and decline in yield made cultivation unprofitable. Nine years ago, when his cultivation had been going well, a bank had motivated him to take a loan for irrigation purposes against which he could get a subsidy of Rs 5,000. He took a loan for Rs 25,000 from the Grameen Bank, Cheeral. Even though he had a sound financial position, he took the loan only because he would get the subsidy amount. Out of the total loan amount sanctioned, he drew only Rs 9,000 as he had own savings with him. The remaining Rs 16,000 he kept as deposit in his bank account. The repayment of the loan was prompt only for three instalments. By 2001, he faced severe financial crisis and could neither repay the loan nor make investments in agriculture due to low production and low prices for his produce. In such a situation he thought of drawing from his deposit in the bank, but he was informed that there was not enough money in his account

(Contd.)

(Box 5.3 contd.)

as the bank had withdrawn the money as repayment of the interest on his loan amount. Joseph was angry that the bank had not taken his prior permission to do this kind of 'adjustment'. Finally, the bank suggested Joseph to take a second loan for Rs 5,000 to repay the initial loan with interest. Joseph again failed to repay the loan due to the crisis in the agricultural sector. Now his outstanding loan has mounted to Rs 100,000.

Source: Nair and Menon (2007).

The case studies cited here, and many more, formed part of the basis for the analysis of the state-specific chapters included in this volume.[1] Case studies of households with incidence of suicides or distress from Maharashtra, Andhra Pradesh, and Kerala's plantation economy of Wayanad, provide us with graphic account of the dimensions of the present pervasive agrarian crisis. There is earnest effort in all these chapters to bring out different dimensions, linkages, and roots of farmer distress, and help in preparing a comprehensive agenda covering all the phases from relief to full restoration of upward mobility.

The following pen picture of a typical farmer struggling to move up in the drought-prone area has been constructed from the studies included in this book and also from the recent literature on farmers' distress. It is in the nature of a synthetic exercise incorporating the important personal features of the farmer and the characteristic features of the local milieu in which he operates.

- He/she carries a heavy accumulated burden of debts to many creditors with none ready to lend any further and all insisting on immediate repayment. The proximate cause is low-negative returns over a successive run of years due to crop losses caused by weather, pests, etc. and crash in prices. Given his/her strong urge to move upwards, he/she persists in borrowing and investing in the hope of having a good season helping him/her to clear all debts and earn handsome returns. This turns out to be a mirage owing to his/her following features.
- His/her transition from subsistence-oriented rainfed agriculture to modernized agriculture is only half way through. His/her traditional support systems with minimal access to markets and

monetized economy are breaking down much before his/her getting stabilized in modernized agriculture. Institutions like PRIs, SHGs, and cooperatives need to be developed and activated to build up an institutional support system meeting the needs of modernized agriculture.

- A first generation farmer entering modernized agriculture with some experience in its intricacies but not fully competent in the skills it needs. Particularly weak in dealing and coping with channels of modernization—markets, traders, input dealers, institutional finance. No effective access to crucial services—insurance, warehousing, post-harvest processing, and export.

- Drought-prone areas have still to adjust to the needs of modernized agriculture. In Karnataka, tank irrigation, which provided a protective cover to traditional subsistence crops, has drastically declined. The government policies have encouraged investment in tube wells, leading to overexploitation of groundwater and diversion to water-intensive crops not suitable for the drought-prone areas. Recently, in Karnataka, there was drought in three successive years wiping out the gains made over several years by using the groundwater. The most serious barrier to growth in the drought-prone areas is the absence of agro-climatic regional planning without which watershed development, adoption of dryland technologies, and choice of crops and rotations best suited to drought-prone conditions would not make any progress.

- The crops to which he/she turns—cotton, groundnut, vegetables, sericulture—have price fluctuations that could make all the difference between losses that could bankrupt and gains that could raise him to the status of a successful farmer. He does not realize that the emerging context of globalization and liberalization raises the probability of his making losses and diminishes that of gains. Nobel-laureate Professor Joseph Stiglitz has recently observed that 'External liberalisation has exposed India to inequalities of the global trading system. The large number of debt-ridden cotton farmers taking their lives across India is clearly related to the American agricultural subsidies that depress prices and make cotton farmers elsewhere worse off' (see 'India Warning on Globalization' by Soutik Biswas, BBC News website, 20 December 2006).

LOOKING AHEAD

The following agenda covers briefly the main lines of action needed to help the upwardly mobile farmer to get out of the unprecedented crisis confronting him.

Relief to Suicide Victims' Families and Prevention of Suicides

- Follow-up on the victim families to check whether the quantum of compensation offered was adequate to relieve distress and rehabilitate the family.
- The family members and friends are most likely to notice the signs when a farmer seriously contemplates suicide. Suicide may be triggered by an event such as threat by a moneylender but the underlying distress would be known to many in the family and in the village. An informal arrangement in which a group of elders in the village talk to such farmers may go some way in preventing suicides. The group can persuade the farmer to approach his creditors for help much before becoming a serious defaulter.
- The lending institution should develop a system of early warning to advise the farmer when he is still in a position to get out of his indebtedness. He may be induced to seek rescheduling of past debts and restrictions may be put on his further borrowing if he does not cooperate.
- Activists and extension agencies should help the farmer to select a portfolio of activities suited to his situation and resource position. He may also be assisted to adopt this portfolio.

Protecting and Maintaining Credit-worthiness of the Farmer

- There should be a cell in the Zilla Parishad to systematically monitor weather conditions, crop prospects, and price behaviour. This would be of some help in mapping the pockets in the district where widespread distress may occur. The panchayats and farmer organizations, such as SHGs and their federations, could be alerted for taking timely action to advise farmers about avoiding risks and taking protective measures.
- A computerized data bank should be developed based on the information card of the individual farmer, which is updated

to reflect his credit status. The credit status should be a code categorizing the farmer from 'Excellent' to 'Not Eligible for Credit'. The list of farmers with their status codes should be periodically sent to the lending institutions. The cases of farmers not eligible for credit should be scrutinized and the farmer should be advised about the steps he should take and a programme worked out to help him.

Services Needed by Modernized Agriculture

- An essential requirement of modernized agriculture is that the farmer should acquire capacity for collective action to undertake activities such as planning of crops, water use, management of pests and diseases, choice of varieties and seeds, and use of fertilizers and manures and to get prompt and reliable services from traders, input suppliers, research and extension agencies, regulated markets, and insurance companies.
- The agriculture department of the state government should closely monitor the status of these services and their delivery mechanisms on the basis of feedback from farmers and their organizations. Similarly, there should be concurrent evaluation of programmes and policies designed to help the farmer, such as price support, in getting adjusted to modernized farming.

Adjustment to Globalization

The case of cotton indicates the distress caused by wild fluctuations in international prices. The international markets tend to have many imperfections such as gaps in information and non-competitive structures. Even more important, developed countries with very small proportions of population and GDP in agriculture can give large and liberal subsidies to their farmers. They also can and do use many devious ways to make it difficult for developing countries such as India to increase their agricultural exports, including value-added products. The Government of India, the state governments, and farmers' organizations will have to remain vigilant to ensure that agricultural exports and imports are on terms that are fair to the farmer and the country.

Progressive Modernization of Agriculture

The support to the upwardly mobile farmer should be part of a broader policy regime to promote modernization benefiting all sections of farmers and consumers. The problem of indebtedness is equally serious among the poor farmers. They do not commit suicide but lead a miserable life trapped in indebtedness to moneylenders. The media does not pay much attention to them, which makes it all the more necessary to have them in the data bank mentioned above and to have adequate safety nets for them to lift them above the threshold where they either get a productive livelihood outside agriculture or become upwardly mobile in agriculture. Landless labourers should also remain in the agenda, and the need to raise them to the threshold of a productive livelihood should receive as much priority as the needs of the poor farmers and the upwardly mobile farmers. In the coming years, the policy maker would be under severe pressure from two opposing sources—rising aspirations of those neglected so far but who are emerging as a force in the political arena. The opposing forces are the pressures for liberalization and globalization from the powerful block of developed countries. Simultaneously, the policy maker has also to face serious environmental concerns and the huge burden of dysfunctional lobbies created by over two decades of regressive subsidies and the loan *mela* culture. His policy making record so far has been poor. But he remains the only hope in a grim situation.

NOTE

1. The four state-specific chapters included in this volume are based on secondary data, household surveys, focus group discussions, as well as case studies of farming households. The analysis benefited from all these sources but the case studies are not included as part of the text of these chapters because of space constraints.

REFERENCES

Mishra, S. (2006a), *Suicide of Farmers in Maharashtra*, http://www.igidr.ac.in/suicide/suicide.htm, Report submitted to the Government of Maharashtra, Mumbai: Indira Gandhi Institute of Development Research.

NSSO (2005), *Situation Assessment Survey of Farmers: Some Aspects of Farming*, NSS 59th Round (January–December 2003), Report No.

496 (59/33/3), New Delhi: Ministry of Statistics and Programme Implementation, Government of India.

Nair, K. N. and V. Menon (2007), 'Distress, Debt and Suicides among Agrarian Households: Findings from Three Village Studies in Kerala', Working Paper No. 397, Thiruvananthapuram: Centre for Development Studies.

Revathi, E. (2007), 'Farmers' Suicides in Andhra Pradesh: Issues and Policy Concerns', mimeo, Hyderabad: Centre for Economic and Social Studies.

6

Agrarian Distress and Farmers' Suicides in Maharashtra[1]

SRIJIT MISHRA

INTRODUCTION

The dynamism exhibited by the Indian economy with the unprecedented high rate of growth after the initiation of economic reforms, especially after the mid-1990s, has several perplexing factors. The impressive growth is largely a story of the service sector and to a lesser extent for industry whereas agriculture is lagging behind. There is deceleration in agriculture, which is bordering on distress. This has manifested in suicides of farmers in many parts of the country. Maharashtra is one of the examples where on the one hand, there has been high incidence of farmers' suicides and, on the other hand there has been high growth in the non-agricultural sector, resulting in further polarization of urban-rural disparities.

The main objective of this chapter is to examine the policy and other factors contributing to agrarian distress in the state. The issue is examined within the context of reforms and the prevailing conditions of agriculture in relatively resource-poor and high-risk zones of the state. Besides making use of secondary sources, a fairly large household survey was conducted in Western Vidarbha,[2] where high incidence of farmers' suicides has been witnessed.[3] While the selection of the suicide households was from the list of reported cases, the choice of the control group household involved selecting someone similar to the suicide household based on socio-economic characteristics such as land size, caste, or other parameters. These were supplemented through focus group discussions and some village level information.

Besides this brief introduction, the chapter is divided into five other sections. Beginning with the contrasting growth between agricultural and non-agricultural sectors, the second section also shows the shrinking share of agriculture in the state income with a continued dependence of the majority of the workforce in the sector. It presents a picture of agriculture in the state with particular reference to Western Vidarbha. The third section examines changes in the policy domain and the institutional arrangements in the context of certain identifiable risks with agriculture. The fourth and fifth sections are detailed analysis of farmers' suicides with particular reference to Western Vidarbha. Based on the primary survey, it tries to identify some important socio-economic correlates of farmers' suicides. The sixth section is a brief analysis of the policy response to the larger crisis. The last section presents certain concluding observations.

AGRARIAN CONDITION IN MAHARASHTRA

Maharashtra is among the richest states of India. In 2004–5, its per capita net state domestic product (NSDP) of Rs 32,170 in current prices was second only to Haryana, yet its head count ratio of poverty at 31 per cent in 2004–5 is higher than the all-India average of 28 per cent.[4] One of the main reasons for the incidence of high poverty amidst high level of average income in Maharashtra is due to the extreme sectoral and regional inequalities. The sectoral inequality worsened in the post-reform period because of near stagnation of agriculture. Table 6.1 shows sectoral growth rates during the pre-reform (1980–1 to 1992–3) and post-reform (1993–4 to 2004–5) periods. While overall growth rates in the industrial sector have shown deceleration, it is agriculture that reached the lowest level, from 3.5 per cent in the pre-reform period to 0.8 per cent in the post-reform period. The dismal growth performance of agriculture caused further contraction of its share in gross state domestic product (GSDP) from 19.5 per cent in 1993–4 to 11.3 per cent in 2003–4 (Figure 6.1). What is more, during the same period the proportion of rural households depending on agriculture declined only marginally, from 73 per cent to 67.6 per cent as per National Sample Survey (NSS) estimates. At the turn of the century, 55 per cent of all workers and 80 per cent of rural workers in Maharashtra

Table 6.1
Sectoral Linear Trend Growth Rates in Maharashtra
during the 1980s and 1990s

Sector	1980–1 to 1992–3 (1980–1 prices)	1993–4 to 2004–5 (1993–4 prices)
Agriculture	3.5*	0.8
Industry	6.2*	3.4*
Services	7.4*	7.5*
Total GSDP	6.1*	5.2*

Note: GSDP indicates Gross State Domestic Product. * indicates that the linear trend growth rate is significantly different from zero at 95 per cent confidence interval.
Source: Calculated from State Domestic Product (State series), *http://mospi.nic.in/* (accessed on 2 October 2007).

were still dependent on agriculture. Thus, it is not just a coincidence that this is also the period that witnessed an increase in the incidence of farmers' suicides—a symptom of the larger agrarian crisis.

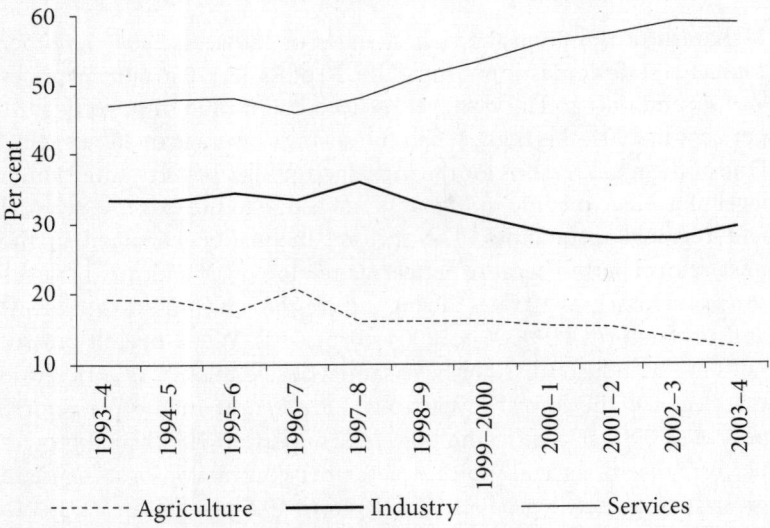

Figure 6.1: Sectoral Distribution of Gross State Domestic Product at Current Prices: 1993–4 to 2003–4

Source: Central Statistical Organisaton, State Domestic Product (State series), *http://mospi.nic.in/mospi_cso_rept_pubn.htm* (accessed 30 September 2007).

Table 6.2
Some Agrarian Features across NSS Regions of Maharashtra

	Konkan	Western Maharashtra	Northern Maharashtra	Marathwada	Western Vidarbha	Eastern Vidarbha	Maharashtra
Average rainfall (May–October), 55+ years (mm)	3,116.9	798.0	825.3	749.8	833.5	1,267.8	1,169.8
Area irrigated (NIA/NSA), 1998–9 (%)	8.2	25.5	16.4	15.3	8.9	33.8	17.4
Area under Cereals, TE 2004–5 (%)	90.0	71.7	50.4	45.5	18.6	65.7	48.5
Per capita cereal, TE 2004–5 (Kg)	174.5	133.0	173.1	203.9	102.1	162.6	155.4
Average land size, 2003 (hectares)	0.8	1.4	2.0	2.5	2.6	1.6	1.9
Gini coefficient of land possessed, 2003 (%)	46.3	50.1	48.8	45.0	42.1	46.1	49.8
Small & marginal farmers, 2003 (%)	92.3	79.1	68.2	56.5	49.7	69.7	68.6
Agricultural labourers/cultivators, 2001 (%)	54.4	54.2	116.7	88.2	182.7	130.0	91.6
Expenditure/Agricultural Receipts, 2002–3 (%)	44.6	44.8	43.6	38.7	52.8	63.6	45.6
Returns per farmer household, 2002–3 (Rs)	3,746.8	12,474.6	17,318.1	18,338.6	13,277.8	4,812.1	3,746.8
ST, SC, and OBC farmers, 2003 (%)	59.4	31.1	79.1	40.6	86.6	94.5	56.8

(Contd)

(Table 6.2 contd.)

	Konkan	Western Maharashtra	Northern Maharashtra	Marathwada	Western Vidarbha	Eastern Vidarbha	Maharashtra
Farmers not liking farming, 2003 (%)	44.9	39.7	34.4	48.6	29.9	33.6	39.3
SMR for male farmers, 2001–4	27.0	33.4	43.9	49.3	110.2	45.2	49.8

Note: NIA and NSA denote Net Irrigated Area and Net Sown Area respectively; TE denotes Triennium ending; ST, SC, and OBC denote Scheduled Tribe, Scheduled Caste, and Other Backward Classes, respectively; and SMR denotes Suicide Mortality Rate, i.e., suicide deaths per 100,000 persons. Average rainfall is for the period 21–44 weeks (May–October) for more than 55 years. Per capita cereal production for TE 2004–5 is calculated with 2001 census of rural population.

Source: For average rainfall, *http://agri.mah.nic.in/* (12 October 2007); for area irrigated, *http://www.indiastat.com* (accessed 10 October 2007); for area and production, personal communication, Commissioner Agriculture, Government of Maharashtra; for population related data, Census of India; for suicide data, personal communication, CID, Government of Maharashtra; all other estimates are based on calculations using unit level data from the 33rd schedule, 59th round of National Sample Survey on Situation Assessment Survey of Farmers, 2003.

Agriculture in Maharashtra is not only stagnant but is characterized by extreme and growing regional disparities, with parts of the state experiencing negative growth during the last decade,[5] resulting in extreme forms of distress among the farming community. For administrative purposes, Maharashtra is divided into six divisions. The National Sample Survey also divides this into six regions based on agro-climatic conditions. The two classifications are broadly similar but for a few deviations.[6] In the subsequent discussion, use is made of the NSS regions.

Some agrarian features across the six NSS regions are given in Table 6.2. For Western Vidarbha, average rainfall during May–October is considered somewhat better at about 35–80 millimetres more than the drought-prone regions of Western Maharashtra and Marathwada, but is much lower than the rain-assured Eastern Vidarbha or the rain-abundant Konkan region. Area irrigated in Western Vidarbha is also among the lowest. This along with other infrastructural and public intervention backlogs leading to regional imbalances, which were pointed out by the Dandekar committee (Government of Maharashtra [GoM] 1984), has continued and remains a grievance of the region. The region's advantage has been its black soil, which is considered good for cotton, and as a result, relatively larger areas have been under this cash crop. In recent years, there have been many difficulties with cotton cultivation and there has been a shift away from this crop (Figure 6.2). What is worrying is that the area under cereals, which has been relatively lower, has also been declining. Using the 2001 rural population as base, the per-capita cereal production in this region is among the lowest.

The average land size possessed per farmer household is higher in Western Vidarbha and this is also reflected in a less inequitable land distribution and a relatively lower proportion of small and marginal farmers. These seem to match with characteristics of relatively land-abundant rainfed agriculture where productivity is lower. It is perplexing to know that this is also a region with relatively higher proportion of agricultural labourers. This is indicative of the fact that there is a divide between the land-owning cultivating class and the landless agricultural labourers. Another matter of concern is that expenditure, as a proportion of receipts, is relatively higher in agriculture. Part of this could be because of the high costs

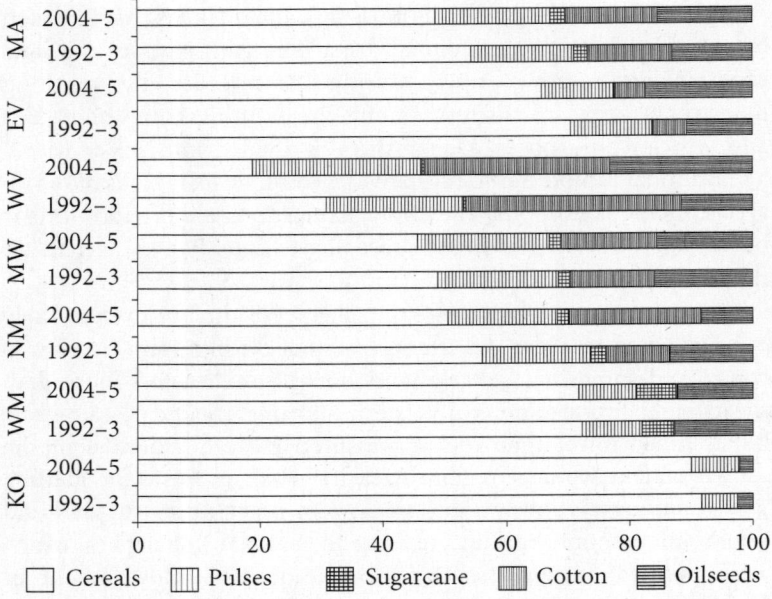

Figure 6.2: Cropping Pattern Across Regions of Maharashtra: TE 1992–3
and TE 2004–5

Note: TE denotes Triennium ending, KO denotes Konkan, WM denotes Western
Maharashtra, NM denotes Northern Maharashtra, MW denotes Marathwada,
WV denotes Western Vidarbha, EV denotes Eastern Vidarbha and MA denotes
Maharashtra.

Source: For TE 1992–3, *Districtwise Agricultural Statistical Information of Maharashtra,
Part-II*, 1999, Office of the Commissioner Agriculture, Government of Maharashtra;
for TE 2004–5 personal communication, Commissioner Agriculture, Government of
Maharashtra.

of cultivation. Then again there was the drought in 2002–3, but its
severity was much more in Konkan, Eastern Vidarbha, and Western
Maharashtra (GOM 2003). After controlling for these, the returns to
cultivation in Western Vidarbha are poor. With such returns, farmers
are not able to adequately meet even the requirements of food. This
is reflected in the fact that while poverty is declining (particularly
among farmers), the same is not true with regard to the number of
undernourished (Reddy and Mishra 2008).

The relatively larger proportion of farmer households that are either scheduled groups or other backward classes (largely comprising Kunbis, a peasant community, and nomadic tribes) and the virtual absence of the politically dominant caste, such as Maratha farmers in Western Maharashtra, have worked to the disadvantage of Vidarbha. Despite all odds, the Western Vidarbha farmer likes his profession—a relatively lower proportion of the farmers do not like farming. This is not enough to save the Western Vidarbha farmer. The region is under a severe agrarian crisis and its worst manifestation is in the high incidence of farmers' suicides.

The plight of the Western Vidarbha farmer is in some sense linked with that of cotton or rather its declining profitability because of increasing costs and relatively lower yields. In 2005–6, the Commission for Agricultural Costs and Prices (CACP) projected the costs of cultivation in Maharashtra per quintal of cotton at Rs 2303.[7] The minimum support price (MSP) fixed by the government was Rs 1760 for short staple and Rs 1980 for long staple. The farmer receives the MSP if the produce is of fair average quality and if it is sold in authorized centres. In practice, farmers sell the produce in the open market where the prevailing price could be lower than the MSP.[8]

In 2004–5, production of cotton was a record high worldwide, as also in India. Maharashtra's production at 52 lakh bales showed a 68 per cent increase over the previous year's 31 lakh bales.[9] This was largely because of a record yield of 297 kilograms/hectare in the state. Regardless of this, Maharashtra's productivity at 64 per cent of the national average continues to be among the lowest. This growth would have bypassed parts of Western Vidarbha where the monsoon scenario in May–October 2004 was largely deficient. In short, the cotton farmer in this region faced both price as well as yield shocks simultaneously.

On cotton prices, there are a number of other relevant factors. Excess international supply at a lower price is also because of direct and indirect subsidies leading to dumping by the USA. During the period 1998–2003, cotton export prices from USA were lower than their cost of production by more than 50 per cent on average (Murphy et al. 2005). Domestic policies in India have led to the removal of quantitative restrictions and subsequently reduction of import tariff from 35 per cent in 2001–2 to 5 per cent in 2002–3. All these exposed the domestic prices to the volatility of international prices that has been adversely affecting the cotton farmer. Similarly, excessive cotton exports leading to an increase in yarn prices can adversely affect the handloom and power loom weavers. The dismantling of the monopoly cotton procurement scheme has meant that when the farmer is being exposed to the global market, there is no mechanism that will guard him/her against the price volatility.

The Western Vidarbha farmer seems to be at the receiving end because of the uncertainties from weather, market shocks, and apathy of public policies leading to grave consequences, as indicated by the alarmingly high incidence of farmers' suicides. The next section elaborates on the role of the state.

PUBLIC INVESTMENT, CREDIT, AND RELATED ISSUES

The state's role in agricultural growth is examined by looking at, among others, public investment in agriculture, rural financial market, and agricultural extension. That there was gross neglect of agriculture by the state in the 1990s is reflected from the drastic decline in capital expenditure on agriculture and allied activities (Table 6.3). Only as a response to the crisis that erupted in the increasing numbers of farmers' suicides was there a revival of capital expenditure in the initial years of the Tenth Plan (2002–5). This is also evident when one normalizes the real capital expenditure with per hectare of net sown area or as a proportion of agricultural NSDP. The actual impact of the

Table 6.3
Real Capital Expenditure (RCE) on Agriculture and Allied Activities in Maharashtra at 1993–4 Prices

State	Eighth Plan 1993–4 to 1996–7	Ninth Plan 1997–8 to 2001–2	Tenth Plan 2002–3 to 2004–5
RCE on agriculture and allied activities (Rs)	1,276	970	2,616
RCE per hectare of net sown area (Rs)	710	548	1,484
RCE as per cent of agricultural NSDP (%)	6.4	5.2	15.1

Note: NSDP denotes net state domestic product.
Source: Chand, Chapter 2, this volume.

Table 6.4
Backlog in Irrigation Sector across Regions in Maharashtra

Region	As on			
	30 June 1982	01 April 1994	01 April 2000	01 April 2002
Vidarbha	38.1	55.4	59.8	62.2
Marathwada	22.9	32.4	32.9	33.1
Rest of Maharashtra	39.1	12.6	7.4	4.7
Total	100.0	100.0	100.0	100.0

Source: GoI (2006).

recent revival would be known after some time lag, say, four-to-five years. Nevertheless, an analysis of the nature and pattern of real capital expenditure would be worthwhile. Based on regional imbalances in development within the state and the expenditure required to bridge this, sector-wise backlogs were calculated by the Dandekar committee (GOM 1984). Over time, the irrigation sector backlog has increased for Vidarbha (Table 6.4). Further, expenditure on irrigation for the period 2002–3 to 2004–5 indicates a shortfall from the outlay to the tune of Rs 2528 crore for Vidarbha, Rs 1148 crore for Marathwada, whereas there was excess expenditure for the rest of Maharashtra to the tune of Rs 1586 crore (Government of India (GOI) 2006).

Recent trends from basic statistical returns (BSR) provided by the Reserve Bank of India with regard to agricultural credit in Maharashtra from 1991 to 2004 indicate the following. Credit utilization to agriculture as a proportion of total credit utilization in the state has declined from 20.2 per cent to 11.2 per cent—this is largely offset by an increase in personal loans. Agricultural credit utilization is shifting from rural regions to urban areas, with Mumbai's share having increased from 5 per cent to 48 per cent. Within agriculture, the share of direct finance reduced from 79 per cent to 51 per cent. Even after excluding Mumbai, the region-wise distribution shows a decline in the share of both direct and indirect finance components of agricultural credit in Vidarbha—for the seven districts of Western Vidarbha agricultural credit (both direct and indirect) declined from 18.0 per cent in TE 1992–3 to 14.3 per cent in TE 2003–4.

The *Situation Assessment Survey of Farmers* (33rd schedule, NSS 59th round, 2003) indicates that in Western Vidarbha, as compared to Maharashtra, a relatively higher proportion of farmer households are indebted, whereas average outstanding amount per farmer household is lower (Table 6.5). Across caste groups and size-class of land possessed, one also observes a positive association with proportion of the indebted and the average amount of outstanding debt per farmer household. When we consider the amount outstanding per indebted

Table 6.5
Farmer Indebtedness: 2003

Household Identification	Western Vidarbha		Maharashtra		India	
	Amount outstanding per farmer HH (Rs)	Indebted HH, %	Amount outstanding per farmer HH (Rs)	Indebted HH, %	Amount outstanding per farmer HH (Rs)	Indebted HH, %
Scheduled Tribes	7,887	43.6	6,379	36.1	5,506	36.3
Scheduled Castes	9,538	53.6	8,845	46.5	7,167	50.2
Other Backward Castes	16,892	68.2	18,205	58.1	13,489	51.4
Other castes	19,060	53.5	21,417	60.5	18,118	49.4
Near landless farmers	4,514	33.7	12,263	44.1	7,663	47.5
Marginal farmers	4,813	40.9	7,563	44.8	7,511	44.6
Small farmers	11,431	54.7	16,421	55.8	13,714	50.9
Semi-medium farmers	12,109	67.6	18,229	64.1	23,169	57.5
Medium farmers	23,929	75.7	38,400	73.4	42,139	65.4
Large farmers	99,847	80.6	122,795	83.7	71,763	65.8
All farmer households	14,483	59.9	16,973	54.8	12,585	48.6

Note: HH indicates household. Near landless (0–0.099 hectares), Marginal (0.1–1 hectares), Small (1.01–2 hectares), Semi-medium (2.01–4 hectares), Medium (4.01–10 hectares), Large (10+ hectares).
Source: Estimates based on unit level data from 33rd schedule, 59th round of National Sample Survey on Situation Assessment Survey of Farmers, 2003.

household, then the indebted small farmers have a higher burden than indebted semi-medium farmers. A similar picture emerges for the near landless, but then this group is less than one per cent of the farmer population.

The formal sources account for more than 80 per cent of the amount outstanding in the state. In particular, cooperative banks have been an important source of credit (more than half of the amount outstanding compared to about one-fifth for India) for agricultural purposes in rural areas. However, much of these loans are likely to be outstanding debts, not current loans. A recent study of Yavatmal indicates that more than half the members are defaulters with their credit lines being chocked up from one to many years (Sarangi 2004). This is so because over the years, the cooperative credit institutions were faced with a number of problems—high interest rate, accounting practices were not rationalized, and no professional management to mention a few (Government of India 2004). Per hectare loan in Western Vidarbha districts is relatively lower (Shah 2006).

An important issue raised in our focus group discussions (FGDs) conducted in 98 villages with an average participant size of 6–7 (minimum–2, maximum–9) is that current operational loans are likely to be from moneylenders. In 70 per cent of the FGDs, the availability of the informal loans in the village was mentioned. One participant's remark during an FGD will elucidate the socio-economic dominance of the moneylender. The participant said that: 'Gentleman, you will go away after this discussion. It is we who have to stay in the village. Please do not probe further into the details. Further revelation by us will make our stay in the village difficult.'

Informal loan transactions could be in *dedhi*, under which the debtor has to return the loan around harvest, that is, within four to six months of borrowing, and pay Rs 150 for loan of Rs 100. Similarly, there is *sawai*, involving repayment of Rs 125 for loan of Rs 100. Another popular form of loans for agricultural and social purposes is at an interest rate of Rs 5 to Rs 10 per month for loans of Rs 100. Non-payment of loan leads to rewriting of a fresh loan with some additional credit being given during the start of the next agricultural season.

A conventional form of collateral is land. Creditors now consider it risky because suicides can lead to cancellation of such contracts, and hence, insist on sale of land with a verbal (not legal) promise that it will be sold back to the debtor after the loan is repaid. If required, legal registration expenses on both counts are borne by the debtor. Land seizure or mortgage was mentioned in 17 per cent of FGDs.

Some of the moneylenders would also be traders. Loan taken could be for purchase of an input and repayment through sale of produce. Interlocking of credit, input and output markets are not necessarily enforced by the trader–moneylender, but operating with a single trader–moneylender would save transaction costs to the farmer.

Nearly 50 per cent of FGDs discussed about the paucity of water. Despite delay and deficient rain, there were instances of people (whole villages) opting for a second or third sowing without any groundwater dependence. With seed replacement being almost complete, it contributed to additional expenses for seed. In the last 5–10 years, there has been an increase in the number of spraying for insecticides/ pesticides and an increase in the need and cost for fertilizers. All these added to the cost. The issue of spurious quality of inputs also came up in the discussions. This brings forth some important points. First, the absence of an extension service that could have advised the farmers against late sowing or improper use of other inputs. Second, with new technology on the anvil, there is deskilling and farmers' experience becomes redundant. The private traders selling farm inputs advises the farmer on extension service, leading to supplier-induced demand.

We also came across villages from where many people have migrated out in search of jobs. In 14 per cent of the FGDs, difficulty in getting employment or availability of work at low wages was mentioned. These not only indicate the unavailability of non-farm jobs in the study regions, but also indicate poor public interventions.

One of the important social welfare measures in the state is the Maharashtra Employment Guarantee Scheme (MEGS), which has been in operation since the 1970s. In Western Vidarbha, the share of MEGS expenditure from the state's total MEGS expenditure decreased from 14.4 per cent in 2000–1 to 6.0 per cent in 2003–4. This is much lower than the region's share of rural poor (23 per cent as per 1999–2000 estimates) or its share of rural population (17 per cent as per the 2001 census). The regions share of item-wise expenditure during 2000–1 to 2003–4 was 4.2 per cent for agriculture, 6.5 per cent for irrigation, and 13.0 per cent for horticulture. In particular, it reflects poor intervention in works associated with developing agriculture, either directly or indirectly through interventions in irrigation and horticulture. These observations are in line with the findings mentioned in a recent study that MEGS has been successful as a relief measure largely concentrated in drought-prone areas of Marathwada and Western Maharashtra divisions of the state and has had a limited success as a poverty eradication measure (Vatsa 2005).

FARMERS' SUICIDES IN MAHARASHTRA

Recent Trends and Issues

In 2001, Maharashtra constituted about 9.3 per cent of the all-India cultivator population but accounted for 22.7 per cent of the total farmer suicide deaths in the country during 2001–6. Between 1995 and 2006 the total number of farmer suicides in Maharashtra increased by four-fold from 1083 in 1995 to 4453 in 2006. The increase was largely because of an increase in male farmer suicides. During this period, male farmer suicides as a proportion of total male suicides in Maharashtra increased from 14 per cent to 39 per cent. Suicide

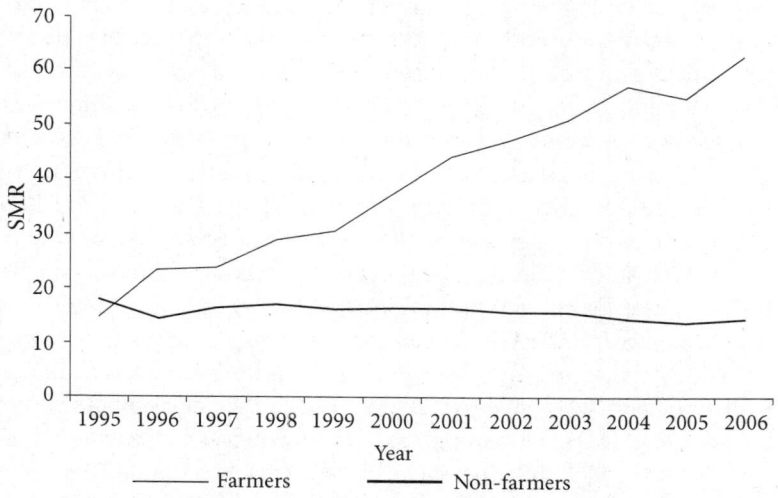

Figure 6.3: Suicide Mortality Rate (SMR) for Males (Farmers and Non-farmers) in Maharashtra: 1995–2006

Source: For suicides, National Crime Records Bureau (various years); for population, Census of India, 1991 and 2001.

mortality rate (SMR, suicide deaths for 100,000 persons) for male farmers nearly quadrupled from 14.7 in 1995 to 62.6 in 2006 whereas SMR for male non-farmers decreased from 18.0 in 1995 to 14.4 in 2006 (Figure 6.3). As indicated earlier, during 2001–4, SMR for male farmers at 110 in Western Vidarbha is among the highest across all regions of Maharashtra (also see Figure 6.4).

The higher incidence of farmers' suicides in Western Vidarbha has been receiving a lot of media attention, more so by the vernacular local dailies. It first came into prominence in the late 1990s. In response, there was a report by the Government of Maharashtra (1998). Some of the reasons identified in the report are the crop failures of 1996–7 and 1997–8, poor returns from cultivation, indebtedness, and inability to service debt among others. One of the first scholarly works focusing on two districts of Western Vidarbha (Amravati and Yavatmal) covering suicides in the late 1990s was that by Mohanty (2001). It provides a historical backdrop to the agrarian changes and links the increase in land ownership among lower castes to the post-independence land ceiling laws. However, these small landowners faced greater losses during the above-mentioned two successive crop failures. Many could not recover costs and these had negative implication on their food-basket, which in a cash crop–intensive cultivation was dependent on returns from cultivation. Size-class wise analysis indicates that crop loss and indebtedness were more important for small farmers whereas for medium and large farmers, family problems, old age and illness, and loss in business and other economic activities also assume importance. A further study in the three districts of Amravati, Wardha, and Yavatmal covering cases in 2002 was taken up by Mohanty and Shroff (2004). Comparing suicide cases with control households, it observes that loss of agricultural income that is linked with adverse weather, market imperfections, and consequently indebtedness along with other social pressures have pushed the farmers to distress. Using evidence from the above two studies, Mohanty (2005) analyses the suicides in a Durkheimian perspective and suggests that lower and middle caste smallholder peasants found themselves trapped between enhanced aspirations and the neoliberal reality of rising debt and declining income, whereas large and medium farmers belonging to the higher castes were faced by failures in business, trade, and politics. He opines that farmers' suicides are an effect of individualization under rapid economic growth where the rural producers find themselves estranged from the traditional agrarian communities. The cotton farmer has lost his competitiveness and profit incomes have declined to significantly negative levels (Mitra and Shroff, 2007). The increasing incidence of farmers' suicides, Lott (2006) indicates, is also because of neglect of agriculture by the state where the people dependent on it seem to have lost their political voice and weight.

The rising suicides among farmers became the concern of the public and the subject matter of public interest litigation. The Bombay High Court directive led to the Tata Institute of Social Sciences coming up with a report (Dandekar et al. 2005) which identified repeated crop failure, inability to meet the increasing cost of input-intensive cultivation, and indebtedness as important factors leading to distress. The report pointed out a larger agrarian crisis and the need for strengthening various support structures among others. The Government of Maharashtra sought independent opinion from the Indira Gandhi Institute of Development Research (Mishra 2006a); this will be elaborated in the next section on micro level analysis of farmers' suicides, where suicide has been considered as symptomatic of the larger agrarian crisis. The Planning Commission constituted a fact-finding team, which reiterated the neglect of Vidarbha with the continuation of backlog in irrigation as well as other infrastructure and the larger agrarian crisis, particularly in cotton cultivation (GOI 2006).

Yet another report (Meeta & Rajivlochan 2006) identified two common factors, viz., one, a feeling of helplessness as farmers were not able to resolve problems and dilemmas, particularly because of lack of funds for various activities (such as health and daughter's marriage) or for repaying loans, and second, the absence of any person, group, or institution to whom they could turn for advice (for agricultural operations, fund related or personal reasons). Such an interpretation does not take away the issues raised by Dandeker et al. (2005) as well as Mishra (2006a). They reiterate that the individual level factors operate in a larger socio-economic context, which includes among other things the failure of the state in agriculture, healthcare, and education. Further, many of the factors identified by them (health, marriage, and agrarian pressure) would be associated with indebtedness. In addition, Mishra (2006a), as will be seen below, has pointed out that suicide is associated with multiple risk factors (indebtedness being only one of them, like suicides it is perhaps an important symptom of the larger socio-economic malaise) that can co-exist and aggravate each other. By investigating some of the households that were already surveyed in Mishra (2006a); Meeta and Rajivlochan (2006) identify additional risk factors, including greater loan burden (and also point out a discrepancy in caste name), and implicitly substantiate the proposition of multiple risks. An important observation by Meeta and Rajivlochan is 'the absence of adequate implementing mechanism' and the need to ensure that many of the existing measures work with greater efficiency.

The public policy concern has resulted in initiatives by the central and state governments. A few studies have highlighted farmers' suicides and the plight of farming in some *talukas* of Nagpur district (Fadnavis et al. 2006) and Marathwada (Kurulkar 2006), as these regions have been exempted from the above-mentioned initiatives. A recent documentary (c. 2007) *Haya Janmavar...* in Marathi highlights the larger situation based on cases from Jalgaon district. There is also a Hindi movie, *Summer 2007* (released in June 2008), where the poor socio-economic condition and farmers' suicides in Maharashtra come to the fore when circumstances force some medical students to their rural services.

It is true that the high incidence of farmers' suicides in Western Vidarbha is a matter of serious concern. The SMR for male farmers in Western Vidarbha during 2001–4 is 110, which is more than eight times the SMR of 14 for males in India. All other regions have SMR for male farmers lower than the state average (Figure 6.4). However, this is no reason to conclude that the situation is better elsewhere because in all regions SMR for male farmers is higher than SMR for male non-farmers. Besides, if one calculates SMR for male farmers across districts for 2001–4, then those with values greater than the state average are spread across. They include six Western Vidarbha districts of Akola (77), Amravati (144), Buldhana (191), Nagpur (86), Wardha (78), and Yavatmal (93) as also the districts of Aurangabad (74), Beed (64), Chandrapur (89), Hingoli (51), Jalgaon (89), Nanded (56), Prabhani (68), and Satara (53).[10] The district of Washim in Western Vidarbha is not included because in the data provided, a large proportion of suicides are classified under the profession of

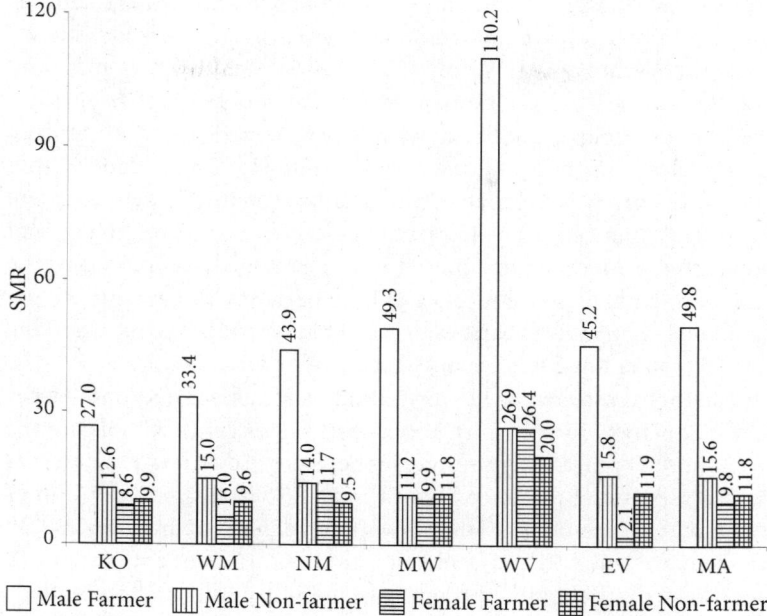

Figure 6.4: Suicide Mortality Rate (SMR) for Farmers and Non-Farmers by Sex
across Regions of Maharashtra: 2001–4

Note: KO denotes Konkan, WM denotes Western Maharashtra, NM denotes Northern
Maharashtra, MW denotes Marathwada, WV denotes Western Vidarbha, EV denotes
Eastern Vidarbha, and MA denotes Maharashtra.

Source: For suicide, personal communication, CID, Government of Maharashtra; for
population, Census of India.

'others'. This is likely to be a misclassification because in a primarily
agricultural district if they are not farmers (or cultivators), then they
are likely to be agricultural labourers and that should be considered
even more critical. Having indicated the larger concern, the focus of
the current exercise including the micro level analysis in the next
section, as indicated earlier, is on Western Vidarbha.

Micro Level Analysis of Farmers' Suicides

To understand the nature and characteristics of the farmers committing
suicide and to unravel the factors that contribute to such an extreme
step, a household survey was conducted in the Western Vidarbha

region. The villages were identified based on a list of farmers' suicides for 2004 and a few cases for January 2005.[11] Data from 111 suicide cases (one Amravati, 21 Wardha, 29 Washim, and 60 Yavatmal) and 106 non-suicide control households spread across 105 villages form the basis of our analysis. In our sample from suicide cases, 91 per cent were males, 55 per cent in the age group of 31–50 years, and 80 per cent were currently married. On educational status, 21 per cent were illiterate, 14 per cent were literate but below primary level, 26 per cent completed primary education but were below matriculation, and the rest were with higher education. On experience in farming, 24 per cent had less than five years, 18 per cent had about 6–10 years, and the rest 58 per cent had more than 10 years of experience.

The three castes with the most number of cases are Kunbi (31 per cent), Banjaras (16 per cent), and Baudh (9 per cent). Kunbis are the predominant peasant community in the selected districts and Banjaras have a substantial presence in Yavatmal and Karanja sub-division of Washim. The size-class of land shows that 14 per cent are marginal, 39 per cent are small, 21 per cent semi-medium, 15 per cent medium, 4 per cent are large, and 7 per cent have not given information on their land ownership.

Risk Factors

Suicide is the complex interplay of multiple factors. A number of risk factors can co-exist and one particular individual can come across all or none of the risk factors identified by us through the household survey. In our sample, the minimum number of risk factors is two and the maximum is nine. The most common thing was indebtedness (86 per cent). From all those indebted, 44 per cent were harassed for repayment of loan, and in 33 per cent of cases the creditor insisted on immediate repayment

Next in importance is fall in economic position (74 per cent). Indebtedness *per se* will not lead to a fall in the economic position, but if it reaches a stage that will lead to sale of assets then it can be associated with a fall in the economic position. Similarly, a fall in economic position can also lead to greater reliance on credit, and thereby, increasing the debt burden. In 55 per cent of the cases, it was observed that the individual concerned had not discussed his/her problem with other family members. He/she was shouldering

Table 6.6
Risk Factors Identified with Deceased Individual

Risk factors	N=111	%
Was the deceased indebted?	96	86.5
Did his economic status deteriorate before the incident?	82	73.9
Did the deceased not share problems with other family members?	61	55.0
Was there a crop failure?	45	40.5
Was there a change in his social position before the incident?	40	36.0
Did the deceased have a daughter/sister of marriageable age?	38	34.2
Was there any suicide occurrence in the nearby villages recently?	36	32.4
Did the deceased have any addictions?	31	27.9
Was there a change in the deceased's behaviour before the incident?	29	26.1
Did the deceased have disputes with neighbours or others?	26	23.4
Did the deceased have some health problem?	23	20.7
Did any death occur in the family recently before the incident?	11	9.9
Has there been any suicide previously in the family?	7	6.3
Are some other family members chronically ill/handicapped?	4	3.6
Average number of risk factors	4.8	
Minimum number of risk factors	2	
Maximum number of risk factors	9	

Note: N indicates number of households. The risk factors are not mutually exclusive, and hence, will not add up to 100 per cent.
Source: Field Survey, Mishra (2006a).

the entire burden that was troubling him and was not sharing the difficulties with others. An avenue for letting out one's pent up feelings and frustration was closed.

Crop failure is mentioned in 40 per cent of the cases and most of these also mentioned about loss in the second or third sowing due to delay in rainfall. A few cases mentioned fire or theft. Crop loss can also happen due to excessive untimely rain, say, during the time of harvest. Crop failure can lead to economic downfall and make it difficult to repay existing loans. This will also increase the need for additional credit. Crop failure leading to fall in economic position is quite straightforward, but the causal links can also be the other way round. A household faced with downfall in economic position or with greater debt burden could not take additional loans for investing in agriculture (say, during a pest attack) and this can lead to a reduction in yield or total crop failure.

Change in social status was identified in 36 per cent of the cases. This can be associated with a fall in economic position. Harassment by creditors or their agents due to non-payment of loans can also lead to social disgrace. Crop failure due to unsuccessful experimentation by a farmer who was recognized as a successful entrepreneur may find a change in his social status—people who earlier came to him for advice are now providing solace to him.

A socially important role of a brother/father is to get one's sister/ daughter married. Communities have norms in terms of age and expenditure.[12] A farmer is largely dependent on a good return from his produce to fulfil this obligation. Thus, crop failure, greater credit burden, or a fall in his economic position can come in his way of fulfilling this obligation. Inability to conduct a sister's/daughter's marriage can be socially humiliating. It can also increase intra-household conflicts. To complete this social obligation, a farmer may also take loans, thinking that he can repay the amount after the harvest. Recent marriage of a sister/daughter or inability to get one's sister/daughter married has been identified as a risk factor in 34 per cent of the cases. We have also taken note of the imitation effect.[13] In some cases (28 per cent), in addition to crop failure, addiction to alcohol also served as a precipitous factor. Under conditions of stress, it is said, an intoxicated individual may indulge in an act of self-harm without being aware of the consequences.

Change in the individual's behaviour was identified in 26 per cent of the cases. These are symptoms and indicate that the individual needs some psychosocial help. Dispute with neighbours/others in the villages was identified in 24 per cent of the cases. This could be related with property disputes or an altercation leading to a social humiliation. Or, it could be a part of his changed behaviour, indicating that he needs some help.

Personal health problem of the deceased was identified in 21 per cent of the cases. From these, 26 per cent (6 cases) were those who were perceived by others with some mental health problem. Illness gets aggravated due to poor economic condition because it makes care seeking difficult. Similarly, ill health can lead to a loan to meet medical expenses and also reduce the ability to work, thus aggravating the economic condition. If the sick person is some other member (3 per cent of the cases), then the breadwinner has the added frustration and helplessness in not being able to provide appropriate care for an ailing parent/spouse/child. Death of another member in the family before the incident was identified in 10 per cent of the cases. The death of the near ones could have been because of not receiving appropriate health care. Inability to provide care is largely because of the poor economic condition rooted in the larger agrarian crisis. Suicide history in the family could be identified in six per cent of the cases. This could be indicative of a genetic factor that increases the chances of suicide as there is setback to agriculture in such households.

In 79 per cent of the cases, suicide was committed by consuming insecticides. These proportions are higher than that indicated for the overall population. This is so because of its easy accessibility in farming households (particularly those cultivating cotton). A hospital that can treat emergencies like poisoning is on an average more than 20 kilometres away. This means that the time taken to reach a treatment centre in these hilly regions can easily be more than an hour. This delay can prove fatal. Restriction on availability and toxic content of pesticides and access to early treatment are important policy parameters. Important policy lessons can be taken from Sri Lanka's experience (Gunnell & Eddleston 2003).[14]

Analysis of a few case studies does indicate that they are enterprising farmers who wanted to contribute to the growth process (Case A and B in Box 5.1 in Chapter 5 of this volume) or wanted to be part of the larger economic activity through his children (Case C in Box 5.1, Chapter 5 of this volume). All the three case studies do reiterate that the risk factors are not mutually exclusive. They can co-exist, they can be interrelated, they can feed into each other, and they can also aggravate each other.

Comparing Suicide Cases with Controls

Comparing suicide cases with control households may help to evaluate the extent of debt burden (Table 6.7). It will also help to identify some characteristics that differentiate the two groups of households. Comparing suicide cases with control households shows that the average outstanding debt is higher in the former by 3.5 times and after normalizing for family size or land size, it is higher by three times, and all these differences are statistically significant. As compared with control households, the suicide cases have, on an average, a lower proportion that owns bullocks (a productive and liquid asset), a lower value of produce, and a relatively greater family size (particularly female members). The relevance of bullocks to the agrarian economy as a productive asset in Indian agriculture is well known (Vaidyanthan 1988, among others). Bullocks are the major means of ploughing, an act that depends on rain and has to be done within a short span before sowing. Hiring of bullocks or tractors will increase costs and the latter may not be as effective in these dryland rain-dependent conditions. Bullocks are also used as a liquid asset that is sold under distress conditions. Thus, absence of bullocks may be a reflection of hardship that the household has been facing. The

Table 6.7
Comparing Suicide Cases with Controls

Average characteristic	Suicide cases		Controls	
	Value	N	Value	N
Debt per household*	38,444	101	10,910	95
Debt per household per person*	7,224	101	2,405	95
Debt per household per acre*	7,079	94	2,365	90
Own bullocks (%)*	42.9	105	63.5	104
Value of produce (Rs 000)	22.9	93	41.4	90
Value of produce per acre (Rs 000)	3.4	93	6.5	90
Family size#	5.53	111	5.08	106

Note: N=number of households. *Difference between suicide cases and controls is statistically significant at 95 per cent confidence interval (CI). # Includes the deceased individual in suicide cases.
Source: Field survey, Mishra (2006a).

suicide case households also had lower ownership of other livestock, agricultural implements, consumer durables, and lower access to basic amenities. These get further compounded with a crop failure or poor yield, leading to lower value of produce. Higher family size would mean that for the same absolute income, per capita expenditure would be reduced. If son-preference would mean that those going for a third child are those with two daughters, then households with more members are also those with more female members. The social obligation and associated expenditure of such households would indicate a higher burden.

In further comparison of the debt scenario, we restrict ourselves to 87 pairs of suicide cases and control households from which we have credit related information. There are 80 suicide cases with outstanding debt from 138 transactions and 49 control households with outstanding debt from 64 transactions. The average number of transactions in suicide cases (1.7) is higher than that in the households control (1.3). Analysis of source of loan indicates greater reliance of

Table 6.8
Average Outstanding Debt per Transaction across Source

Source	Suicide cases		Controls	
	Amount	No.	Amount	No.
Commercial bank	95,000	4	11,500	1
Rural bank	12,200	6	40,800	4
Cooperative bank	32,871	44	14,319	24
Moneylender	23,038	52	12,709	23
Relatives/Friends	16,761	23	–	–
Self-help group	4,500	2	14,000	1
Trader	5,000	1	20,000	1
Landlord/Employer	2,000	1	10,000	1
Others	12,000	4	16,089	9
Not available	5,000	1	–	–
Total	25,739	138	15,617	64

Note: N indicates the number of transactions with outstanding debt. The transactions are from 80 suicide cases and 49 non-suicide controls.
Source: Field survey, Mishra (2006a).

co-operatives in the formal sector and moneylenders in the informal sector (Table 6.8). The reliance on moneylenders and friends/relatives is higher for suicide cases (54 per cent of 138 transactions) than controls (36 per cent of 64 transactions). A very high amount is indicated for suicide case under commercial bank because of a large farmer (owning 28 acres) having an outstanding loan of Rs 2.5 lakh, which was incurred for marriage in the family (in fact, the individual had taken a loan of Rs 5 lakh and had already returned Rs 2.5 lakh). After excluding this extreme case, the distribution of total outstanding debt indicates that 44 per cent is from cooperative banks, 36 per cent from moneylenders, and 12 per cent from relatives/friends. In control households, after excluding a loan transaction with outstanding debt of Rs 98,200 from a rural bank, the distribution of total outstanding debt indicates that 38 per cent is from cooperative bank, 32 per cent is from moneylenders, 16 per cent from other unspecified informal sources, and 15 per cent from self-help groups.

The purpose of credit for outstanding debt by source is given in Table 6.9. After excluding the transactions where purpose is not available, 70 per cent of transactions in the suicide cases and 89 per cent of transactions in the non-suicide controls are for agricultural purposes only. This proportion further increases if we take into consideration transactions from formal sources only (88 per cent in suicide cases and 96 per cent in non-suicide controls). Next to agriculture is marriage, which is mostly from informal sources. For each specific purpose the number of transactions with outstanding debt and the average outstanding debt per transaction is higher in suicide case households when compared with non-suicide control households. The average amount of outstanding debt per transaction for agricultural purposes is greater than Rs 10,000. For marriage, after excluding an extreme observation with outstanding debt of Rs 250,000 from a commercial bank by a suicide case household, the gap is greater than Rs 6000. There is an instance in suicide case households where loan for health expenditure was to the tune of Rs 1.5 lakh. From total outstanding debt (including those where purpose is not available), agriculture being the sole purpose accounts for 64 per cent of the outstanding debt in suicide case households and 79 per cent in control households. Marriage accounts for 19 per cent of the total outstanding debt in suicide cases (reduces to 13 per cent if we exclude the extreme case of

Table 6.9
Average Outstanding Debt per Transaction by Source across Purpose of Loan

Purpose	Suicide cases						Controls					
	Formal	N	Informal	N	Total	N	Formal	N	Informal	N	Total	N
Agriculture	31,695	42	21,311	45	26,324	87	18,934	25	12,704	25	15,819	50
Marriage	250,000	1	31,107	14	45,700	15	–	–	25,000	4	25,000	4
Others	42,261	5	12,765	17	19,468	22	5,000	1	5,000	1	5,000	2
NA	17,833	6	5,125	8	10,571	14	13,333	3	11,700	5	12,313	8

Notes: N indicates number of transactions. NA indicates not available.
Source: Field survey, Mishra (2006a).

Rs 2.5 lakh outstanding debt) and 10 per cent of the total outstanding debt in controls.

The proportion of outstanding debt that is more than one year old (incurred in 2004 or earlier) is 69 per cent in suicide cases and 59 per cent in controls. Overall, 53 per cent of the outstanding debt in suicide cases and 52 per cent of outstanding debt in controls are from formal sources, but for 2004, the most recent year, only 30 per cent of the outstanding debt in suicide cases is from formal sources. In fact, the number of transactions with outstanding debt in suicide cases has been increasing over the years. In 2005, all the transactions reported (for before the start of the agricultural season, as the survey was conducted during March/April 2005) are from informal sources. In suicide cases, reliance on informal sources seems to have increased in recent years.

A statistical exercise is done to compare case-control households. Households' suicide status is a binary dependent variable, Y; 1=case and 0=control. The independent variables, X_i's, are outstanding debt in rupees, a yes/no binary variable on ownership of bullocks, family size, value of produce in rupees, and value of produce per acre of land owned in rupees. Using these, we estimate a step-wise logistic regression,[15]

$$\ln [p/(1-p)] = \alpha + \beta_i X_i + u; \ i = 1, \ldots, 5.$$

where *ln* is natural logarithm, *p* is probability of obtaining a suicide case household, *ln[p/(1−p)]* is the log odds ratio of a suicide case household, α is a coefficient on the constant term, β_i's are the coefficients of the five independent variables, X_i's, and *u* is error term.

While discussing results, instead of coefficients, odds ratio, $e^{\beta i}$, are given because the interpretation is more intuitive—for a unit increase in the independent variable there would be a corresponding change in the odds ratio (probability of a suicide case/probability of a control).

The result for complete case-control analysis of the 68 pairs of observation is as follows. It gives outstanding debt and absence of bullocks as statistically significant variables that differentiate suicide cases from non-suicide control households (Table 6.10). It suggests that if outstanding debt increases by Rs 1000, then the odds that the household is one with a suicide victim increases by 6 per cent and if the household owns bullocks, then the odds that it is a household with a suicide victim decreases by 65 per cent.

If we restrict the case-control pairs to similar land size (the land size of control household in the village not differing from the suicide case household by more than 25 per cent (i.e., if the suicide case household has 4 acres then the control household can have land size in the range of 3–5 acres), then we estimate for 55 pairs of observations. It shows that only outstanding debt per acre of land is a statistically significant variable that differentiates suicide cases from controls. It indicates that

Table 6.10
Results (Odds Ratio) of Step-wise Logistic Regression Analysis

	Complete case-control analysis	Similar land size	Same sub-caste	Similar land size and same sub-caste
N	136	110	32	80
Debt	1.000061			
	(0.0000138)			
	[0.000]			
Own bullocks	0.3462934		0.2092665	0.2156863
	(0.1403603)		(0.1139936)	(0.1258042)
	[0.009]		[0.004]	[0.009]
Debt per acre		1.000325		
		(0.0000776)		
		[0.000]		
Family size			1..352608	
			(0.2021914)	
			[0.043]	
Log likelihood	–74.6497	–61.682649	–42.619212	–35.079024
LR Chi square	39.24	29.13	11.80	7.47
Probably >Chi square	0.0000	0.0000	0.0027	0.0063
Pseudo R square	0.2081	0.1910	0.1216	0.093

Note: Round brackets give standard error, square brackets give probability > |z|. The variables are indicated in the order in which they were selected in the step-wise logistic regression.

Source: Field survey, Mishra (2006a).

if outstanding debt per acre of land owned increased by Rs 1000, then the odds that the household is one with a suicide victim increases by 33 per cent.

If we restrict to the case-control pairs to the same sub-caste, then we have 35 pairs of observations. Our estimation indicates that ownership of bullocks and family size are statistically significant variables that differentiate suicide cases from controls. It suggests that if the household owns bullocks, then the odds that it is a household with a suicide victim decreases by 79 per cent and if the family size increases by one member, then the odds that the household is one with a suicide victim increases by 35 per cent. When we restrict for both land size owned and same sub-caste, then ownership of bullocks turns out to be a significant variable that differentiates the suicide cases from controls. Under other restrictions, even value of produce also turns out to be statistically significant.

Public Policy Initiatives

To mitigate the distress of farmers, two packages, one by the Government of Maharashtra through its Vasantrao Naik Sheti Swavlamban Mission, and the other by the Government of India through the Prime Minister's special rehabilitation measures are being implemented in six of the seven Western Vidarbha districts.[16] The Prime Minister's package includes credit and non-credit components. The credit component involves waiver of interest and rescheduling of overdue loans along with one-year moratorium and provision for fresh credit. The non-credit component aims at reviving the livelihood base of the distressed farmers through programmes of irrigation, watershed development, strengthening of extension service, provision for supply of quality seeds, agricultural diversification towards horticulture and linking its production with agro-and-fruit processing, and the development of allied and non-farm activities. There is provision for committees at the Central and state level for coordination and supervision and creation of appropriate institutional structures at the community level to ensure effective delivery of the package and optimum utilization of the resources in a time-bound manner.

The package is comprehensive in terms of coverage and problems addressed. But, it suffers from some deficiencies in design and implementation (GOI 2007). These can be classified into three categories. First, the design of some of the schemes is not based on the felt needs of households. The state government also conducted a house-to-house census of 17.6 lakh farmer households in the six districts, and identified distress on account of crop failure (more than two-thirds, 69.5 per cent), indebtedness (more than half, 50.4 per cent), social obligations like marriage (17.2 per cent), and chronic disease in the family (5.2 per cent), among others. For organic farming, there was demand from 7.5 lakh households, but only about 40,000 could be included as part of 860 organic farming groups. About 10.6 per cent of the households were in favour of community marriage, but interventions by the government facilitating this could meet less than 5 per cent of the demand (6783 couples were married at a cost of Rs 7.46 crore). More than 9 lakh households were identified with distress due to chronic disease but till 10 January 2007 only 2604 families received Rs 2.6 crore as part of expenditure under ex-gratia assistance from the Prime Minister's National Relief Fund at the disposal of the district collectors (GOM 2007).

Second, the measures are universal in nature, and region or household level specificity is missing. For instance, some schemes, like providing 500 check dams per year in each district, bringing 15,000 hectares under watershed for each district, and Rs 50 lakh ex-gratia per district, are indifferent to the differences in area, population, or agro-climatic conditions. As a result, Yavatmal and Washim get the same allocation though the former is double the size of the latter in terms of area as well as population.

Third, absence of proper institutional arrangements creates problems in implementation and monitoring. There is no information on the impact of the scheme on the people. Monitoring of physical targets and mid-term evaluations are urgently required. Such studies may consider the impact of various schemes on raising incomes, from farm and non-farm sources, and on agricultural productivity, the efficacy of the credit-related interventions, and more importantly to suggest whether the interventions are moving in the right direction with regard to ameliorating distress or there is need for mid-course correction.

The Government of Maharashtra provides for a compensation of Rs 1 lakh if the cause of suicide is because of agricultural crisis. The bereavement is linked to the crisis based on: (i) the deceased being a farmer and its proof lied in the ownership of land; (ii) the deceased being indebted when the incident took place; and (iii) indebtedness is the cause of suicide. Checking legal ownership of land should be considered at the family level, including extended family when land ownership continues to be in a single name, but there are no guidelines to consider the cases of tenants. On indebtedness, it is easy to get proof from formal sources, but difficult to get the details from informal sources; particularly, after the state initiated legal measures against moneylenders, which have driven informal credit underground. After the announcement of the Prime Minister's package with rescheduling of loan and a one-year moratorium, indebtedness from formal sources as a cause for suicide is being dismissed. Thus, it has become much more difficult to conclude that indebtedness happens to be the cause of suicide. It is true that the Government has reviewed old cases and more bereaved families have been made eligible for receiving compensation, but then there is no provision for any follow-up assessment of these households. For more recent cases (mid-2006 to July 2008), the authorities have continued to use the barometer of minimizing inclusion error (not include those who do not deserve) and as a result exclusion error (excluding genuine cases) seems to have increased. From a welfare perspective, any compensation criteria should minimize both the errors, but exclusion error should be considered more serious.

To prove the efficacy of the newly introduced public policy interventions, it is unfortunate that the number of eligible cases of suicides is being used as a measuring rod. Such an approach takes away the discussion from the larger agrarian crisis and limits it to a symptom. The government's own house-to-house census speaks of the magnitude of the distress. The situation is much more serious than it seems and requires efforts that reach large masses at lesser costs.

CONCLUDING REMARKS

An increasing concern in recent years is that the relatively near double-digit growth in the economy has not been inclusive. There is incontrovertible evidence that the rural areas, and in that especially the agricultural sector, has been lagging behind. In the race for growth, Maharashtra with one of the highest per capita income has also benefited, but such growth is concentrated in four urban districts of Greater Mumbai, Nagpur, Pune, and Thane that account for more than half of the state's income. There is hardly any impact of the growth on other regions. Sector-wise, the share of agriculture is shrinking, but a large proportion of the population continues to be dependent on agriculture. Returns from cultivation are on the wane and particularly hard hit have been the cotton farmers of Western Vidarbha. The liberalization of trade and reduced tariffs has particularly gone against cotton farmers. The domestic price support measures too failed to help. Some of the reasons are: high subsidies by the USA leading to price distortions, low import tariffs in India, and failure of the MCPS in Maharashtra. Withdrawal of the state is evident from declining public investment in agriculture, poor government agricultural extension services, and diminishing role of formal institutions in rural financial markets. The farmer now depends on the input dealer for advice leading to supplier-induced demand and on informal sources of credit with greater interest burden. To add to this, 2004 was a rain-deficient year that affected yield in at least some pockets of the selected districts, but overall macro supply scenario being good, the market prices were low. The farmer was exposed to yield and price shocks simultaneously. In short, the systemic risk factors indicate a larger socio-economic and agrarian crisis. A symptom of this is the high incidence of farmers' suicides.

With the macro-level policies and support systems turning unfavourable to agriculture, any one of many of the following micro-level factors can trigger farmers to resort to the extreme step of ending their own life. These micro-level factors that cause vulnerability includes indebtedness, deterioration of economic status, conflict with other members in the family, crop failure, decline in social position, burden of daughter's/sister's marriage, suicide in a nearby village, addictions, change in the behaviour of the deceased, dispute with neighbours/others, health problem, a recent death in the family, history of suicide in family, and other family members being ill. Comparing suicide cases with non-suicide controls, one observes that on an average the former has higher outstanding debt, relatively lower ownership of assets particularly, bullocks a productive and liquid asset, and lower access to basic amenities, a higher family size with more dependants, and a lower value of produce. These indicate that the idiosyncratic factors do not occur in isolation—they are exacerbated because of the larger socio-economic and agrarian crisis.

The policy implication from the above discussion calls for an emphasis on the larger agrarian crisis. Availability of affordable credit requires revitalization of the rural credit market. Risk management should address yield, price, credit, income, or weather-related uncertainties among others. Improving water availability will facilitate diversification of cropping pattern, but this should go hand in hand with policies that increase non-farm employment. Improving agricultural extension that addresses deskilling because of technological changes and also facilitates appropriate technical know-how for alternative forms of cultivation such as organic farming will be of help. There is a strong case for regulating private credit and input markets. Organizing farmers through a federation of self-help groups (SHGs) with government, banks, and other stakeholders playing a pro-active role would be welcome. Interventions should go beyond agriculture to the social sector with an emphasis on education and public health. Besides public institutions, there is need for a greater involvement from the civil society.

NOTES

1. An earlier version was presented in a seminar on 'Agrarian Crisis in India' held at Indira Gandhi Institute of Development Research (IGIDR) on 30 December 2006. Comments from participants as well as referees and discussants, S. Galab, R. M. Mohan Rao, B. B. Mohanty, and K. Vatsa, and discussions with R. Radhakrishna, V. M. Rao, and D. Narasimha Reddy helped in revising the chapter. However, none of them are responsible for the views expressed in this.

2. Western Vidarbha is the Inland Eastern region of the state as per the National Sample Survey (NSS) classification. It includes all the districts of Amravati administrative division (Akola, Amravati, Buldhana, Washim, and Yavatmal) and two districts from Nagpur administrative division (Nagpur and Wardha).

3. From an analytical perspective, suicide is a complex and multifaceted phenomenon. The risk factors can either be in the neurobiological (Mann 2002) or socio-economic (Durkheim 2002) domain. The former are internal to the individual and can be considered as predisposing factors whereas the latter are external in nature and identified as the precipitating factors.

4. For a recent discussion on growth and poverty in Maharashtra, see Mishra and Panda (2006). For Maharashtra's agricultural development till the early 1990s see Sawant et al. (1999), and for a recent discussion see Kalamkar

(2003). For a discussion of some Western Vidarbha districts, see Kulkarni and Deshpande (2006).

5. This is inferred from the primary sector growth rates, which is largely agriculture and allied activities, across administrative divisions calculated for the period 1993–94 to 2002–3 by Shaban (2006).

6. The NSS Inland Eastern region of the state includes all the districts of Amravati division and also Nagpur and Wardha districts from Nagpur division; the remaining districts from Nagpur division constitute the NSS Eastern region. Both these regions/divisions together are referred to as Vidarbha. For our purpose, as indicated earlier, we refer to Inland Eastern region as Western Vidarbha and Eastern region as Eastern Vidarbha. The second deviation is that the NSS Inland Western region includes all the districts of Pune division and also Ahmednagar district from Nashik division; the remaining districts from Nashik division constitute the NSS Inland Northern region. For our purpose we refer to Inland Western as Western Maharashtra and Inland Northern as Northern Maharashtra. Konkan and Marathwada divisions are Coastal and Inland Central regions respectively.

7. Costs include actual expenses in cash and kind incurred in production by the owner, rent paid for leased in land and imputed value of family labour, interest on value of owned capital assets (excluding land), and rental value of owned land (net of land revenue).

8. Till recently, this gap was somewhat met through an additional advance price of Rs 500 per quintal paid under the monopoly procurement scheme. This, coupled with various rent-seeking activities, led to the scheme making a cumulative loss of more than Rs 5000 crore, leading to its discontinuation from 2005–6.

9. These are estimated by the Cotton Advisory Board and differ from the estimates used by the Government of Maharashtra to calculate the value of output in agriculture.

10. The high incidence of farmer' suicides is indicative of a larger crisis. However, this does not mean that the absence or lower incidence of suicides can be identified with no crisis.

11. Occurrence of suicide, a rare event, was the basis for selecting a village. In the village, we surveyed the suicide case household and a control household, conducted a FGD, and also collected some village level information before moving over to another village. The distance from one village to another, more often than not, was 20–30 kilometres or more.

12. In the selected region, instances of females getting married before 18 years of age are prevalent. Marriage expenditure would depend on the economic position of the household and FGDs indicate it to be around Rs

20,000–Rs 40,000 only for marginal/small farmers. With such a small amount, it is not the social custom of marriage but the poor returns from agriculture that becomes relevant.

13. Suicide contagion also increases the responsibility of media and guidelines by the World Health Organization on suicide reportage would be of help (Mishra 2006b). This is not to deny the important role that media played in highlighting the issue of farmers' suicides. For a media perspective on Vidarbha, see Deshpande (2006) among others. For a discussion on the representation of farmers' suicides in the Indian media, see Vij (2006).

14. A recent study suggests that pesticides are not only agents for suicide, but are also part of the causal pathway (London et al. 2005). Exposure to organophosphorous pesticides can affect the central nervous system, which in turn can lead to depression and subsequently suicide. For a discussion on poisoning cases in Yavatmal Medical College, see Bhatkule (2006).

15. In the step-wise procedure, a variable is added if it increases chi-square significance by 0.05 and it is dropped if it increases chi-square significance by 0.1.

16. The predominantly urban district of Nagpur has been kept out of the packages.

REFERENCES

Bhatkule, P. R. (2006), 'Poisoning Cases in Yavatmal Medical College, July 2004–June 2005', in *Suicide of Farmers in Maharashtra Background Papers*, Report submitted to the Government of Maharashtra, mimeo, Mumbai: Indira Gandhi Institute of Development Research (IGIDR).

Dandekar, A., S. Narawades, R. Rathod, R. Ingle, V. Kulkarni, and Y. D. Sateppa (2005), 'Causes of Farmer Suicides in Maharashtra: An Enquiry', Final Report Submitted to the Mumbai High Court, mimeo, Tuljapur: Tata Institute of Social Sciences, Rural Campus.

Deshpande, V. (2006), 'Farmers' Suicides: A Media Perspective', in *Suicide of Farmers in Maharashtra Background Papers*, Report submitted to the Government of Maharashtra, mimeo, Mumbai: IGIDR.

Durkheim, E. (translated from French by John A. Spaulding and George Simpson) (2002), *Suicide* [1897], London and New York: Routledge Classics.

Fadnavis, M., P. Deshpande, A. Kalmbe, P. Kale, and S. Khadatkar (2006), 'Identifying Reasons for Farmer's Suicide in Nagpur District (with special reference to Narkhed, Katol and Kalameshwar)', mimeo, Nagpur: Mahila Mahavidyalaya.

Government of India (2004), *Draft Final Report of the Task Force on Revival of Cooperative Credit Institutions* (Chairman: A. Vaidyanathan), New Delhi: Ministry of Finance, December.

—— (2006), *Report of Fact Finding Team on Vidarbha: Regional Disparities and Rural Distress in Maharashtra with particular reference to Vidarbha* (Chairperson: Adarsh Misra), New Delhi: Planning Commission.

—— (2007), *Report of the Expert Group on Agricultural Indebtedness* (Chairman: R. Radhakrishna), New Delhi: Ministry of Finance, July.

Government of Maharashtra (1984), *Report of the Fact Finding Committee on Regional Imbalance in Maharashtra* (Chairman: V. M. Dandekar), Bombay: Planning Department, April.

—— (1998), *Suicides of Farmers in Maharashtra: A Socio Economic Survey* (in Marathi), mimeo, Pune: Commissioner Agriculture, August.

—— (2003), *Final Copy of Documentation on The Drought of 2002–03: State Level Efforts, http://mdmu.maharashtra.gov.in/pdf/droughtmgmt/drought02-03.pdf* (accessed 12 October 2007), Mumbai: Revenue and Forests Department.

—— (2007), *Farmers' Suicides in Maharashtra: An Overview,* mimeo, Amravati: Office of Divisional Commissioner.

Gunnell, D. and M. Eddleston (2003), 'Suicide by Intentional Ingestion of Pesticide: A Continuing Tragedy in Developing Countries', *International Journal of Epidemiology*, Vol. 32, No. 6, pp. 902–9.

Kalamkar, S. (2003), 'Agricultural Development and Sources of Output Growth in Maharashtra State', Working Paper No. 5, Pune: Gokhale Institute of Politics and Economics.

Kulkarni, A. P. and V. S. Deshpande (2006), 'Agrarian Scenario in Yavatmal, Washim and Wardha Districts', in *Suicide of Farmers in Maharashtra Background Papers,* Report submitted to the Government of Maharashtra, mimeo, Mumbai: IGIDR.

Kurulkar, R. P. (2006), 'Farmers' Suicides in Marathwada Region of Maharashtra State', mimeo, Aurangabad: Swami Ramanand Teerth Marathwada Research Institute, November.

London, L., A. J. Flisher, C. Wesseling, D. Mergler, and H. Kromhout (2005), 'Suicide and Exposure to Organophosphate Insecticides: Cause or Effect?', *American Journal of Industrial Medicine*, Vol. 47, No. 4, pp. 308–21.

Lott, H. (2006), 'Protest and Suicide of Farmers in India: Challenge to Governance and Potential for Strategic Reform', mimeo, Paper prepared for presentation at the conference on Poverty, Inequality and Judicial Access for Poor in South Asia, New Delhi: Jawaharlal Nehru University, 7–9 November.

Mann, J. (2002), 'A Current Perspective of Suicide and Attempted Suicide', *Annals of Internal Medicine*, Vol. 136, No. 4, pp. 302–11.

Meeta and Rajivlocahan (2006), *Farmers Suicide: Facts and Possible Policy Interventions*, Pune: Yashwantrao Chavan Academy of Development Administration.

Mishra, S. (2006a), *Suicide of Farmers in Maharashtra*, http://www.igidr.ac.in/suicide/suicide.htm, Report Submitted to the Government of Maharashtra, Mumbai: IGIDR.

—— (2006b), 'Farmers' Suicides in Maharashtra: Content Analysis of Media Reports', in *Suicide of Farmers in Maharashtra Background Papers*, Report Submitted to the Government of Maharashtra, mimeo, Mumbai: IGIDR.

—— (2006c), 'Farmers' Suicides in Maharashtra', *Economic and Political Weekly*, Vol. 41, No. 16, pp. 1538–45.

Mishra, S. and M. Panda (2006), 'Growth and Poverty in Maharashtra' , Working Paper No. WP–2006–001, Mumbai: IGIDR.

Mitra, S. and S. Shroff (2007), 'Farmers' Suicides in Maharashtra', *Economic and Political Weekly*, Vol. 42, No. 49, 8 December, pp. 73–7.

Mohanty, B. B. (2001), 'Suicides of Farmers in Maharashtra', *Review of Development and Change*, Vol. 6, No. 2, pp. 146–88.

—— (2005), '"We are Like the Living Dead": Farmer Suicide in Maharashtra, Western India', *Journal of Peasant Studies*, Vol. 32, No. 2, pp. 243–76.

Mohanty, B. B. and S. Shroff (2004), 'Farmer's Suicides in Maharashtra', *Economic and Political Weekly*, Vol. 39, No. 52, pp. 5599–606.

Murphy, S., B. Lilliston, and M. B. Lake (2005), *WTO Agreement on Agriculture: A Decade of Dumping (United States Dumping on Agricultural Markets)*, Minneapolis: Institute for Agriculture and Trade Policy.

Reddy, D. N. and S. Mishra (2008), 'Crisis in Agriculture and Rural Distress in Post-reform India', in R. Radhakrishna (ed.) *India Development Report 2008*, New Delhi: Oxford University Press, pp. 40–53.

Sarangi, U. C. (2004), 'Report on Cropping Pattern Optimisation in the Districts of Yeotmal, Jalna and Nandurbar to Augment Livelihood Opportunities' (Study instituted by IGIDR on behalf of United Nations Development Programme and Government of Maharashtra), mimeo, Mumbai: IGIDR.

Sawant, S. D., B. N. Kulkarni, C. V. Achuthan, and K. J. S. Satyasai (1999), 'Agricultural Development in Maharashtra: Problems and Prospects', Occasional Paper: 7, Mumbai: National Bank for Agriculture and Rural Development.

Shaban, A. (2006), 'Regional Structures, Growth and Convergence of Income in Maharashtra', *Economic and Political Weekly*, Vol. 41, No. 18, pp. 1803–15.

Shah, D. (2006), 'Resurrection of Rural Credit Delivery System in Maharashtra', in *Suicide of Farmers in Maharashtra Background Papers*, Report Submitted to the Government of Maharashtra, mimeo, Mumbai: IGIDR.

Shroff, S. (2006), 'Cotton Sector in Maharashtra', in *Suicide of Farmers in Maharashtra Background Papers*, Report Submitted to the Government of Maharashtra, mimeo, Mumbai: IGIDR.

Vaidyanathan, A. (1988), *Bovine Economy in India*, New Delhi: Oxford and IBH.

Vatsa, K. S. (2005), 'Asset-based Approach to Household Risk Management: An Analysis of Selected Case Studies in South Asia', Ph.D. Dissertation, School of Engineering and Applied Science, George Washington University, Washington D. C.

Vij, R. (2006), *Journalism for Life: A Look at Indian Media Representation of Farmer Suicides*, New Delhi: Centre for Media Studies.

7

Farmers' Suicides and Unfolding Agrarian Crisis in Andhra Pradesh

S. GALAB, E. REVATHI, AND P. PRUDHVIKAR REDDY

INTRODUCTION

There is growing unanimity that Indian agriculture has been passing through a phase of serious crisis. The crisis in agriculture has many manifestations, of which suicides of farmers is one that came as a rude jolt. Andhra Pradesh is the first state that drew countrywide attention through a spate of suicides by farmers. They are only symptoms of the deeper malaise of agrarian crisis, which is a result of a combination of factors that include growing marginalization process in agrarian structure, increasing fragility of land and water resources, and sustainable cropping practices, all of which were aggravated by the neglect of public support systems due to the economic reform process. This chapter analyses the agrarian crisis in Andhra Pradesh in the light of a series of suicides of farmers in the state. The chapter is divided into five sections. The first section analyses the magnitude of the problems and socio-economic characteristics of farmers committing suicide in the state during the period of about the last one decade, based on a review of a number of earlier studies, and a primary survey extending over four select districts. The other sections try to unravel the factors underlying the agrarian crisis in the state. The second section, after drawing attention to the deceleration of agricultural growth, analyses the changes in the agrarian structure, the growing resources fragility, and unsustainable cropping patterns, especially in resource-poor areas. The third section analyses the growing costs and declining returns with particular reference to a few important crops in the state. The fourth section discusses the economic reforms in relation to agriculture, institutional retrogression, growing burden of informal debt, and rural distress. The last section briefly sets out the way to come out of the crisis.

FARMERS' SUICIDES IN ANDHRA PRADESH

There are considerable regional disparities in the agrarian economy of Andhra Pradesh.[1] The early phase of green revolution in the state was largely confined to resource-rich South Coastal Andhra, and bypassed the semi-arid and rainfed areas of Rayalaseema, North and South Telangana, and North Coastal Andhra. In the recent phase of the spread of green revolution, practices took the form of farmers in dry regions adopting high value crops that would integrate them with markets that fetch higher returns from agriculture. Those farmers with low resource base were forced to embark on high-risk investments, including those on groundwater development. The disparities in agriculture also took the form of disparities in vulnerability. The suicides of farmers in the state reflect these regional dimensions as well.

The first spurt in farmers' suicides occurred in the mid-1980s (1986–7). There were a few cases in the later years, so the early warnings of crisis in agriculture in the form of suicides were ignored as aberrations. It is the second spurt, beginning with 1997 that became a recurring phenomenon. Table 7.1 shows that between 1998 and 2006, there were 4403 farmers' suicides in the state. According to a judicial commission appointed by the state government, between 14 May 2004 and 10 November 2005 alone there were 1068 suicides of farmers, and in addition there were 277 starvation deaths of weavers in the same period (Kumar 2005). The commission also brings out the fact that the state was rocked by the revelation that 26 debt-ridden farmers of Guntur district had sold their kidneys.

There have been a number of studies in the state based on sample investigation of households of the victims to analyse the proximate causes for the suicides. A few of them are reviewed here to understand the nature of the stress factors behind the suicides. In 1998, a people's tribunal (RSC 1998) heard the depositions from 60 farming households of victims drawn from across five districts in the state. Of the 60, the

Table 7.1

District-wise Estimates of Suicides of Farmers in Andhra Pradesh: 1998–2006

District	1998	1999	2000	2001	2002	2003	2004	2005	2006	Total
Srikakulam		1				1	10	2		14
Vizianagaram							9			9
Visakhapatnam				1			22	11	2	36
East Godavari			1	2		1	21	3	5	33
West Godavari	1			5	1	2	26	1	2	38
Krishna	4	1	1	3	7	3	52	15	8	94
Guntur	8	22	2	7	4	1	103	27	12	186
Prakasam	1			2	3	1	62	9	2	83
Nellore		3					38	9	5	52
Coastal Andhra										545(12.37)
Chittoor	3			2	3	3	51	9	15	86
Cuddapah				4	8	12	39	9	7	79
Kurnool	4	4	2	4	3	18	102	25	49	211
Anantapur			51	97	78	104	176	25	24	555
Rayalaseema										931(21.14)

(Contd.)

(Table 7.1 contd.)

District	1998	1999	2000	2001	2002	2003	2004	2005	2006	Total
Medak	15	3	2	8	2	4	139	44	39	256
Mahabubnagar	6	21	36	26	41	46	203	63	25	467
Warangal	69	100	98	119	123	92	88	51	26	766
Nizamabad	9	1		11	5	4	122	43	20	215
Adilabad	9	8	5	13	4	13	67	52	27	198
Karimnagar	31	10	6	30	20	16	216	96	32	457
Khammam	20	5	3	6	6	7	49	23	16	135
Nalgonda	5	1	10	11	18	12	142	59	33	291
Ranga Reddy	5		3	6	2	7	58	41	20	142
Telangana										2927(66.47)
Andhra Pradesh	190	180	220	357	328	347	1795	617	369	4403

Source: Compiled by Andhra Pradesh Rythu Sangam (APRS), 2007.

majority (42) reported their dependence on wells or borewells for irrigation, on which they invested a substantial part of their resources. Thirty-three of them reported water shortage as the main reason for their crop failure that triggered the suicides. Of the 21, who reported heavy borrowing for investing on the wells or bore wells or deepening of bore wells, only one reported bank credit, while the rest had borrowed from private informal sources at very high interest rates.

An AWARE (1998) study covered a sample of 92 households of farmers who committed suicide, from across 10 districts. The major cause reported by most of the victims' households is accumulated debt for digging or deepening of the wells. The trigger has been repeated crop losses or in some cases the failure of the bore wells. The resulting inability to repay the loans and the feeling of threatened self-respect acted as the triggers.

A citizen's report (CES 1998) investigated 50 households of deceased farmers in Warangal district of the state. In all the cases, the cause seems to be failure of crops due to inadequate water sources. But 72 per cent of the households depended on own groundwater sources, while 28 per cent depended on tanks, which failed to provide full protection because of monsoon failure. But all of them have invested heavily on cotton crop. The 40 households who borrowed earlier from the institutional sources had to resort to informal credit sources, because they were defaulters due to their inability to clear the outstanding debt from the institutional sources. Because of the increase in the number of borewells and the fast depletion of the groundwater table, there were attempts to repeated deepening of these sources. The study reports a case study of a village in the district, where 30 years ago there were only six tube wells and the water table was at a shallow level of six feet from the surface. But by 1998 there were 1800 bore wells, half of which were dry, and the water table was at 240 feet. Groundwater dependence and the related risks looms large in the emerging agrarian crisis. But there may not be any parallel to the worst scenario than Musapally village in Nalgonda district. For a village with 2000 acres under cultivation, there are more borewells than people, 6000 borewells dug at an estimated investment of Rs 6.52 crores. About 85 per cent of these wells are a failure (Sainath 2004).

The report by Vandana and Jafri (1998) is based on an investigation of 27 victims' families. In all the cases, crop failure and uncleared debt burden were the immediate causes. But the causes for heavy debt in 16 cases were the investment in wells or bore wells, and nine of them reported either dried up bore wells or bore wells with insufficient water supply. As in the study by Sudhakumari (2002), most of the victims are male members in the age group of less than 40 years. With a few exceptions, the majority of them belong to Backward Communities (BCs), Scheduled Castes (SCs), or Scheduled Tribes (STs). Barring the ST households, their educational level indicates that most of them were literate and with some years of schooling.

Another study, covering a few contiguous drought prone districts of Andhra Pradesh and Karnataka (Vidyasagar & Chandra 2004), shows that though droughts are nothing new to the region, the yield increasing technology without development of adequate infrastructure is superimposed on an agrarian structure where land institutions are not rationalized. The adoption of market-driven commercial and mono-cropping culture has led to severe stress on the local resource base. The reforms in the form of declining public investment and declining institutional support in input supply have contributed to rising costs. Trade liberalization under the WTO regime, especially in the case of oilseeds and cotton, has exposed the dryland farmers to the volatility of global prices. All the attempts of the Government of India during the 1980s to boost domestic production of oilseeds turned farmers of some of these districts into mono-crop oilseeds producers, but the liberalized imports that depressed prices spelt the undoing of all their agricultural initiatives. The study notes that 55 per cent of farmers in the state do not get Minimum Support Prices (MSP). The worst affected are the marginal, small, and even medium farmers who do not get the MSP because of their dependence on traders for credit not only at high rates of interest, but also tied up with the purchase of output at prices lower than the MSP.

The study of farmers' suicides by Rao and Suri (2006) focuses on indebtedness, which they observe is not new to rural Andhra Pradesh, but what is new is suicides due to growing indebtedness. Their findings show that growing costs and low returns have been at the bottom of the causes for suicides. Their diagnosis is that these changes in the agrarian situation are due to neoliberal policies that are in turn a consequence of loss of power of the farming community resulting in their removal from the policy-making process.

That the agrarian crisis not only continues but has spread to different regions and to a number of other high-value crops than cotton is evident from yet another field study (Bhushan and Reddy 2004). The study covers 168 cases of farmers' suicides spread over five districts extending to all the three regions of Andhra Pradesh. Only six of the 168 victims were women. Most of the victims (65.63 per cent) are in the prime of their working life (31–50 years). Most of them have school education, and only a small proportion (18.75 per cent) are illiterate. The majority of them (65.63 per cent) belong to the Backward Communities (BCs) but there are also 25 per cent of them belonging to the so-called forward castes. The proportion of suicides among Scheduled Castes (6.21 per cent) and Scheduled Tribes (3.12 per cent) are relatively less than their proportion in the population. Most of them are small-marginal farmers and 22 per cent of them are pure tenants leasing land. But, most of the others, with some land, also leased additional land, and thus 77 per cent of the victims had leased in some land. There was invariably mono-cropping with all being non-food commercial crops, such as cotton (52 per cent), chillies (33 per cent), sunflower (11 per cent), tomato (8 per cent), sugarcane (6 per cent), and mulberry (3 per cent). So, one need not be a cotton farmer to face high risks. It could be any commercial crop in the reform atmosphere that is adequate to put one under stress and vulnerable to suicide. The trigger in these cases again is the failure of water resources. Most of them depended on tube wells (61 per cent), and some of them on dug wells (22 per cent), and tanks (19 per cent). All these sources failed but 53 per cent reported total failure and most of them were tube well owners, while 47 per cent reported inadequate water. This case study yet again emphasizes that reforms that have no sensitivity to the regional specificities of agriculture are likely to put disproportionately high pressure on the livelihood of farmers in dryland areas where there is no canal irrigation and the entire burden of developing water resources through wells or bore wells is on the shoulders of farmers.

In addition, there has been increasing pressure on the farmers in terms of meeting the basic social services such as education and health, which are increasingly privatized and have been emerging as a significant part of domestic expenditure needs. A combination of these stress factors have been at the back of the crisis in the farming sector, which has been manifesting into widespread suicides, particularly from 1997. Some other recent studies are that of Sridhar (2006); Bhat & Kumar (2006), among others.

Table 7.2
Socio-economic Profile of Sample of Deceased Farmers

Characteristics	Warangal	Mahabubnagar	Anantapur	Guntur
Gender Composition (%)				
Male	97.2	95.0	86.0	98.0
Female	2.8	5.0	14.0	2.0
Literacy (%)				
Literates	36.5	23.8	51.9	53.6
Non-literates	63.5	76.2	48.1	46.4
Age Group (%)				
30–50	66.1	64.8	52.9	57.9
Family type (%)				
Nuclear family	93.9	71.4	75	75.4
Joint family	7.1	28.6	25	24.6
Social Group (%)				
SC	15.8	19.1	9.6	13.4
ST	4.4	17.1	1.9	6.2
BC	67.5	52.4	39.4	51.2
OC	12.3	11.4	49.0	27.2
Ownership Category (%)				
Pure tenant	7.5	nil	3.9	8.3
Pure owner	63.5	80.7	70.6	60.9
Owner cum tenant	25.2	18.3	24.5	30.8
Operational Category (%)				
Marginal & small	91.2	69.1	38.2	78.1
Semi-medium & medium	8.5	30.1	50.1	21.8
Big	nil	nil	12.0	nil
Farmers with agriculture as the only occupation (%)	47.5	80.7	82.8	92.3
Compensation received (% to total sample deceased)	71.5	45.5	47.6	50.2

Source: Field study (Revathi 2007).

A recent sample survey of socio-economic characteristics of farmers who committed suicide in 2003 and 2004 in four districts (Revathi 2007) also shows similar results comparable to earlier studies (Table 7.2). The survey also brought out the factors contributing to the incidence of suicides. These findings may be summed up as the following stylized facts:

- The more the dependence on the groundwater sources of irrigation for cultivating non-food crops, the higher the probability of the farmers committing suicides.
- The more the dependence of the farmers on the high-cost informal credit for working capital and fixed capital for raising the non-food crops, the higher the plausibility of the farmers committing suicides.
- The tenant farmers who grow non-food crops and depend on groundwater sources for irrigating the crops are more prone to commit suicides.
- Though suicides are not class specific, the majority of suicides are among small and marginal farmers, followed by upwardly mobile middle level farmers.
- Suicides are noticed among all caste groups of farmers but most of the cases are among 'other backward classes' (OBCs).
- The more the dependency of farmers for the inputs on the markets, and the more the market volatility in output prices, the greater is the probability of farmers committing suicides.
- For farmers whose household income is more diversified, the less is the possibility for them committing suicides.
- The more is the proportion of their income that farmers are forced to spend on education, health, and social events, the more is the probability of farmers committing suicides.

AGRARIAN CONDITIONS IN ANDHRA PRADESH

Marginalization of Holdings and Resource Stress

In Andhra Pradesh, as elsewhere in the country, the growing demographic pressure and to an extent the land reforms have brought about drastic changes towards growing marginalization in the structure of landholdings. By 2000–1, the proportion of marginal and small holdings increased to 83 per cent and the area under these holdings accounted for almost half of the total operated area. The average size of holdings has come down to 1.25 hectares, and the average size of the predominant marginal and small holdings was at a much lower level of 0.7 hectares. (Government of Andhra Pradesh (GOAP) 2005) Though the state has witnessed considerable diversification in agriculture, one disconcerting fact is that there has been a trend decline in the net sown area (NSA) and an increase in the fallow land. The available evidence shows that this has to do to a considerable extent with the agro-climatological factors as well. The NSA in the Coastal Andhra region has marginally increased over the years. But, both in Rayalaseema and Telangana, there has been a decline in the NSA and increase in the fallow land. Part of the explanation is that there has been land degradation to the extent of 19 per cent of the cultivated area in the state, and much of the degradation is associated with growing stress and erosion in the dryland regions, which extend across Rayalaseema and South Telangana (Reddy & Behera 2003). The declining NSA and increasing fallows are seen more as an early warning on the pursuit of unsustainable agricultural practices in the drought prone areas.

Shift to Non-food Crops

Agriculture in Andhra Pradesh, which was traditionally based on food crops accounting for 70 per cent of NSA, witnessed considerable diversification in the cropping pattern since the 1980s. By 2004–5, three crops viz., cotton, groundnut, and rice, emerged as major crops, together accounting for over 50 per cent of the total cropped area (GOAP 2005). And, of course, two of these three crops, groundnut and cotton, are not only vulnerable to agronomic risks but also to the risks of exposure to global market volatility under the WTO regime. Ironically, the market-driven shifts in cropping pattern contributed to more stability in relatively better endowed regions, and aggravation of risks in resource poor areas. While the area under millets declined drastically in all regions, the substitute crops in the resource rich South Coastal Andhra (SCA) were pulses, which have contributed to reduce risk and increase supplementary income to mainly rice growing farmers of the region. But the crops that came as substitutes for millets were groundnut in moisture deficient Rayalaseema and hybrid cotton in both South and North Telangana, which have exposed farmers of these regions to high risks and unstable incomes (Reddy & Reddy 2007).

Unsustainable Trends in Irrigation

Until late into the 1980s, irrigation was a major area of public investment in the state, and by 1990–1 the proportion of net area irrigated in the NSA reached 40 per cent and the surface water sources, viz., tanks (23 per cent) and canals (43 per cent) together, accounted for about two-thirds of the area while about one-third was from groundwater sources. But this has been completely reversed since the early 1990s (GOAP 2005). The disturbing dimension of irrigation development in the state is the phenomenal increase in the dependence on groundwater resources through wells and bore wells. By 2004–5, wells and tube wells emerged as the single largest source of irrigation, accounting for almost half of the net irrigated area. The minor surface irrigation works such as tanks, which are most often centuries old water harvesting structures that provided for not only direct irrigation but also served as conjunctive percolation, augmented

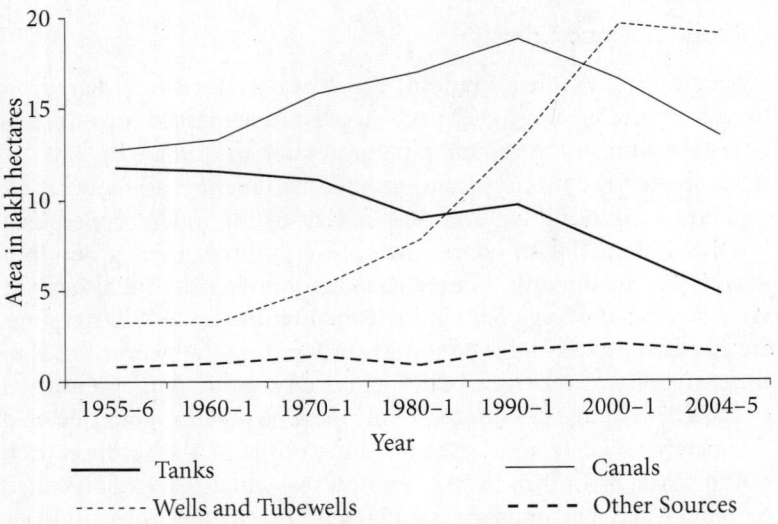

Figure 7.1: Net Irrigated Area by Sources: 1955–6 to 2004–5

Source: Government of Andhra Pradesh (2005).

groundwater sources of irrigation. But the decline in the area under tanks, and disproportionate increase in the groundwater exploitation, implies an emergence of a disjunction that spells an ecological doom, especially for dry regions. The trends in this disjunction are well captured by Figure 7.1.

Groundwater sources in the dry regions of Telangana and Rayalaseema account for 64 and 70 per cent of the total net irrigated area respectively. There is growing unsustainability of the groundwater dependence in the dry regions (GOAP 2005). Rayalaseema and South Telangana, with lowest levels of rainfall, have over the years come to acquire very high levels of dependence on groundwater, relative to development of surface water sources. North Telangana too has come to depend heavily on groundwater resources. This is to a considerable extent due to a fall in public investment in irrigation, leaving farmers in dry regions to bear substantial burden of irrigation provision from their own resources, by heavy borrowing often from non-formal sources, and yet end up with very unstable and low level of groundwater-based irrigation. These three regions with a very high ratio of area irrigated by groundwater to area irrigated by surface water are also the zones of farmers' suicides.

Deceleration in Agricultural Growth

The general impression that Andhra Pradesh is an agriculturally prosperous state is a misleading notion, to say the least. Table 7.3 shows a steep deceleration in the state both in the growth of production and productivity in the 1990s. The agricultural performance slipped much below the national average, leaving the state in a serious crisis. Even leaving aside comparing with international standards, a comparison with states within India, shows that Andhra Pradesh's agricultural performance is not very impressive. Comparison between Andhra Pradesh and all India shows that during the 1990s, Andhra Pradesh's agricultural growth rate was below the national average. During this period, India's agricultural sector grew at an annual average rate of 3.1 per cent compared to 2.2 per cent in Andhra Pradesh. The average annual total factor productivity (TFP) growth rate for all India was 1.3 per cent compared to 0.8 per cent in Andhra Pradesh (World Bank 2003). Studies show that the yields in Tamil Nadu in sugarcane, groundnut, and oilseeds are 1.4, 1.9, and 1.9 times higher respectively,

Table 7.3

Growth Rates of Area, Production, and Yield of Principal Crops in
Andhra Pradesh and All India: TE 1982–3 to TE 1999–2000

Period	State / All India	Area	Production	Yield
TE 1982–3 to TE 1987–8	Andhra Pradesh	−1.32	0.63	1.98
	All India*	−0.39	2.08	2.48
TE 1987–8 to TE 1992–3	Andhra Pradesh	1.53	6.30	4.79
	All India*	0.88	5.00	4.08
TE 1992–3 to TE 1999–2000	Andhra Pradesh	−0.60	1.35	1.96
	All India*	0.36	2.75	2.39

Note: * Weighted average of 15 major states.
Source: Himanshu (2005).

as compared to Andhra Pradesh. Tobacco and pulses in Uttar Pradesh, onion in Gujarat, and cotton in Madhya Pradesh have yields that are 3.9, 1.8, 1.5, and 1.2 times higher than the corresponding yields in Andhra Pradesh. Except chillies, maize, and to a lesser extent cotton, the change in productivity levels is not impressive. Groundnut is the worst case; the productivity shows actual decline after half a century and in the face of the fact that the crop has emerged as the second major crop in the state. In terms of overall yields of foodgrains, Andhra Pradesh continues to lag behind Punjab, Haryana, Tamil Nadu, West Bengal, Uttar Pradesh, and even Kerala (World Bank 2003).

What is disconcerting is that productivity of rice, the single largest crop that accounts for a little over 30 per cent of cropped area, ranks first in 20 out of 22 districts in the state, and contributes about 25 per cent of the state's agricultural GSDP, has slowed down since the early 1990s. While a slowdown in the growth of rice production in the 1990s is a national phenomenon, growth in Andhra Pradesh has been the lowest among the five major rice producing states (World Bank 2003). Andhra Pradesh, particularly in the 1990s, witnessed a steep decline in investment in agriculture. The rate of growth of public investment in agriculture declined from 8.5 per cent in the 1980s to 1.4 per cent in the 1990s. Private investment in agriculture, which increased at a rate of 4.7 per cent in the 1980s, actually turned negative to −3.8 per cent in the 1990s (Dev & Ravi 2003). A World Bank (2003) study offers what it considers as two plausible explanations for the decline in private investment in agriculture—one relating to the change in landholding structure and the other relating to incentive systems. The study considers the declining average size of holdings and the rising tenancy as dampers on the ability of the farmers to invest. On the incentives side, it finds that, in the case of cotton and groundnut, in the wake of import liberalization there has been a downward pressure on the prices, while in the case of rice, it is the regulated market with movement restriction that depresses prices and denies incentive for investment. This is only a partial explanation because the ground reality suggests that while the aggregate private investment shows a declining trend, regionally there are divergent trends in private investment in agriculture. For instance, the widespread prevalence of tenancy in the resource rich Coastal Andhra makes the land owners to siphon off agricultural surpluses in the form of high rents and thus leave very little for the tenants to invest in agriculture. On the other hand, in dryland regions even small–marginal farmers, with relatively low access to irrigation facilities and no surpluses of their own, have been getting into deep debt to augment investments in agriculture in the form of tapping groundwater resources. This kind of forced investment in dryland regions is concealed by the aggregated level of declining private investment.

RISING COSTS AND FALLING RETURNS FROM AGRICULTURE

The data of Cost of Cultivation Scheme (CCS) at the state level clearly indicate a steep increase in the costs of cultivation (at constant prices with base triennium ending 1983–4), especially in the case of paddy and cotton during the 1990s (Table 7.4 and Figure 7.2). For instance, the cost of cultivation of rice per hectare which was Rs 4345 in the early 1990s, increased to Rs 6287 in the mid–1990s, to Rs 6435 in the early years of this decade (early 2000) and to Rs 6235 in 2004–5, recording an increase of 54 per cent; while there was no such corresponding increase in yield. For instance, the increase in the rice yields is around 46 per cent during the same period. The growth in yield rates has either stagnated or has been marginal during the 1990s and early 2000s in the state as a whole. Even the marginal improvements in yield rates during the 1990s are achieved through higher input use, resulting in a faster increase in the cost of cultivation than the value of output. It is also evident that the rate of change in money income over cost C2 (which includes all actual expenses in cash and kind incurred in production by owner, rent paid for leased in land and imputed value of family labour, interest on value of owned capital assets (excluding land) and rental value of owned land (net of land revernue)) from rice is far less in Andhra Pradesh compared to other rice growing states indicating the gravity of the situation in Andhra Pradesh (Reddy and Reddy 2007). The cost per hectare of groundnut was around Rs 2623 during the early 1980s but increased to Rs 3923

Table 7.4

Trends and Indices of Cost-C2 of Important Crops in Andhra Pradesh

[Rupees per hectare (Triennial averages)]

Period	Rice		Groundnut		Cotton	
	Current prices	Constant prices	Current prices	Constant prices	Current prices	Constant prices
Early 1980s	4,054	4,054 (100.0)	2,623	2,623 (100.0)	NA	NA
Mid-1980s	6,196	5,095 (125.7)	3,553	2,922 (111.4)	4,881	4,014 (100.0)
Early 1990s	10,258	4,345 (107.2)	8,845	3,746 (142.8)	12,344*	5,228 (130.2)
Mid-1990s	18,874	6,287 (155.1)	11,216	3,736 (142.4)	19,807	6,598 (164.4)
Early 2000s	26,229	6,419 (158.3)	14,653	3,586 (136.7)	22,418	5,487 (136.7)
2004–5$	26,441	6,235 (153.8)	16,639	3,923 (149.6)	23,710	5,591 (139.3)

Note: Figures in parentheses indicate the indices deflated by CPIAL with triennium ending 1983–4 as base. Early 1980s refers to triennium ending 1983–4; mid-1980s refers to triennium ending 1986–7; early 1990s refers to triennium ending 1993–4 for rice; 1992–3 for groundnut; mid-1990s refers to triennium ending 1996–7 for rice and groundnut; 1994–5 for cotton; early 2000s refers to triennium ending 2002–3 for all the three crops. All these notes hold good for all the tables depicting the yield and returns. It is ideal to use input indices to deflate the costs, but due to data constraints, we have used the CPIAL indices for deflating the costs. These indices are used because human labour and bullock labour costs alone comprise of one third of the cost C2 and the same will be over 50 per cent in operational cost. $ Projections by Government of India; NA indicates not available; * Average of mid-1980s and mid-1990s, as the data for the early 1990s is not available

Sources: Cost of cultivation data collected in Andhra Pradesh and estimated by the Directorate of Economics and Statistics, Ministry of Agriculture and Cooperation, Government of India (GOI), New Delhi as well as by P. Prudhvikar Reddy for his doctoral thesis.

in the year 2004–5, that is, an increase of about 50 per cent (Table 7.4 and Figure 7.2). However, the rate of increase in the yields is far less (36 per cent) compared to costs at constant prices. The increase in costs per hectare indicates that the increased use of inputs but the increase in yields are much less in proportion. By and large, a similar increase in the cost per hectare of cotton at constant prices (an increase of 39 per cent) over the early 1980s is evident. If we take cost C3 which includes the cost of managerial contribution as well, net income accrued to the farmers (in current prices) turns out to be negative since the early 1980s, even for groundnut and cotton. Net income even over cost (C2) in current prices is negative in the case of groundnut. Net income at constant prices showed a decline for all the three crops during the period under consideration (Figure 7.3, 7.4, and 7.5). For instance, the net income of paddy per hectare declined from Rs 370 in the early 1980s to Rs 251 in the year 2004–5 with fluctuations in between. The decline is much steeper in groundnut followed by cotton (Table 7.4). The cost of cultivation data also shows that the growth in the yield was less in Andhra Pradesh compared to other states in the case of groundnut and cotton. For instance,

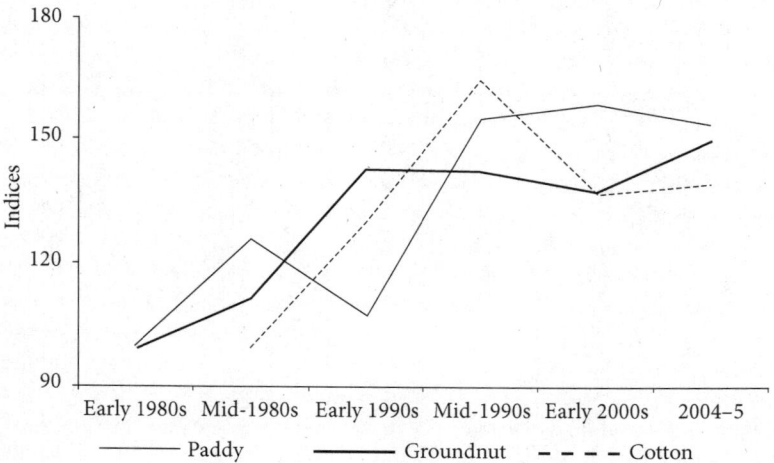

Figure 7.2: Trends in Costs per Hectare at Constant Prices

Source: Cost of cultivation of principal crops in India, Directorate of Economics and Statistics, Department of agriculture and Cooperation, Ministry of Agriculture, Government of India, New Delhi, various years.

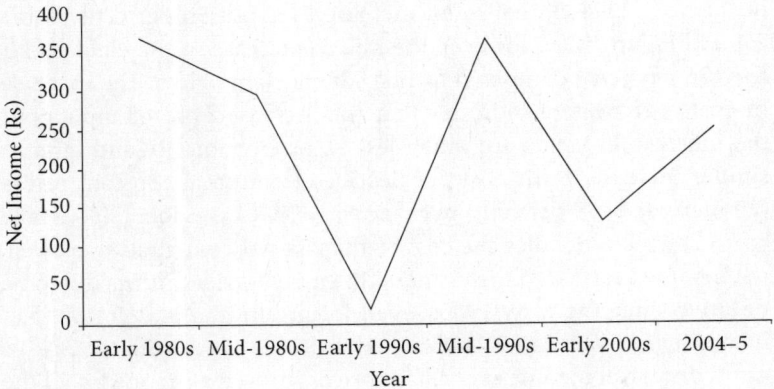

Figure 7.3: Trends in Net Income per Hectare of Rice at Constant Prices

Source: Cost of cultivation of principal crops in India, Directorate of Economics and Statistics, Department of Agriculture and Cooperation, Ministry of Agriculture, Government of India, New Delhi, various years.

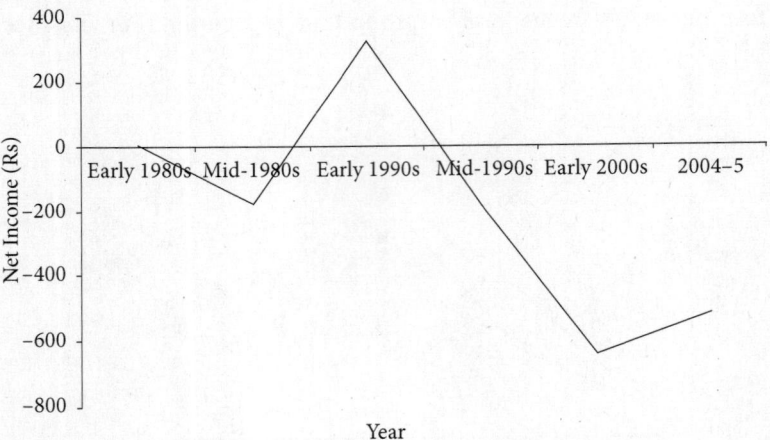

Figure 7.4: Trends in Net Income per Hectare of Groundnuts at Constant Prices

Note: Base is early 1980s based on State Agricultural Domestic Product Implicit Price Deflator.
Source: Cost of cultivation of principal crops in India, Directorate of Economics and Statistics, Department of Agriculture and Cooperation, Ministry of Agriculture, Government of India, New Delhi, various years.

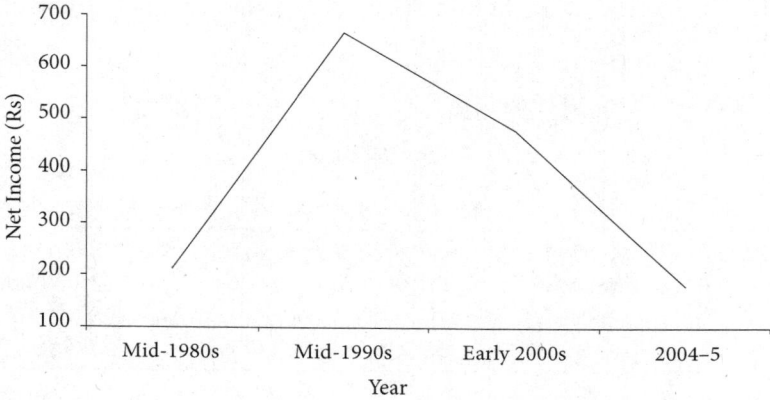

Figure 7.5: Trends in Net Income per Hectare of Cotton at Constant Prices

Note: Base is early 1980s based on State Agricultural Domestic Product Implicit Price Deflator.

Source: Cost of cultivation of principal crops in India, Directorate of Economics and Statistics, Department of Agriculture and Cooperation, Ministry of Agriculture, Government of India, New Delhi, various years.

the change in the rate of growth in yield per hectare of cotton has been negative in Andhra Pradesh (–0.07 per cent) while other states showed a positive trend. Similarly, the change in the rate of growth in yield of groundnut was less (0.50 per cent) in Andhra Pradesh while other states showed a better performance.

Growth differentials in yield rates across the crops under consideration was also observed. It is interesting to examine how these growth differentials in yield are linked to price and trade policies, especially after the early 1990s. For instance, in the absence of any major technological advancement in any of the crops, agricultural policies like price and trade policies may explain the decline or stagnation in yields. However, the viability of farming also depends on other policies related to subsidies, investment, etc. The linkages could be direct and indirect. Direct intervention of support prices (in line with costs) for agriculture produce may ensure profitability in the event of stagnant yield rates. Indirectly, trade policies can be used to maintain the market prices at a certain level, that is, allowing exports or curtailing imports when there is a glut in the market for a certain commodity, and curtailing exports and allowing imports when there is a fall in production. In the Indian context, there is a policy mix of these two types.

Table 7.5

Yield and Returns at Current Prices of Important Crops in Andhra Pradesh

(Figures per hectare)

Period	Yield (quintals)	Gross returns (Rs)	Net income over C2 (Rs)		Net income over C3 (Rs) in current prices
			At current prices	At constant prices*	
Rice					
Early 1980s	32.55	4,423	370	370	−36
Mid-1980s	38.31	6,563	367	298	47
Early 1990s	42.32	10,321	63	22	−963
Mid-1990s	46.63	20,290	1,416	361	−471
Early 2000s	47.47	26,917	689	135	−1934
2004–5$	47.47	27,903	1,462	251	−1182
Groundnut					
Early 1980s	7.08	2,626	3	3	−259
Mid-1980s	7.01	3,331	−222	−180	−577
Early 1990s	10.84	9,776	931	325	46
Mid-1990s	8.80	10,494	−722	−184	−1,844
Early 2000s	8.92	11,359	−3,294	−643	−4,760
2004–5$	9.64	13,635	−3,004	−517	−4,668
Cotton					
Mid-1980s	5.94	5,145	264	215	−180
Mid-1990s	12.04	22,427	2,620	668	639
Early 2000s	12.92	24,887	2,469	482	227
2004–5$	13.35	24,777	1,067	183	−1,304

Note: $ Projections by Government of India; Cost C3 = Cost C2+10 per cent of Cost C2 on account of managerial functions performed by the farmer; *Base: Early 1980s; State agricultural domestic product implicit price deflator.

Source: Cost of cultivation data collected in Andhra Pradesh and estimated by the Directorate of Economics and Statistics, Ministry of Agriculture and Cooperation, Government of India (GoI), New Delhi, as well as by P. Prudhvikar Reddy for his doctoral thesis.

As the prices are the ultimate indicators of the direct and indirect measures, let us examine the trends in output prices and international prices during the last two decades as the output prices are the actual prices received by the farmer for his produce. The evidence shows that despite the reforms and deregulation, minimum support prices have increased at a faster rate during the 1990s when compared to the 1980s for all the important crops except groundnut (Reddy & Reddy 2003). As we have seen earlier, during the period under consideration cost has grown much faster than output prices, indicating pressure on farmer's profit margins. In the case of cotton and to some extent for groundnut, variations in output prices continued to exist after the 1990s, indicating the volatility of profits (Figure 7.6). The volatility is also observed in the international prices, especially in cotton and to some extent in groundnut, indicating the uncertainty of the returns in growing these two crops (Figure 7.7). We also made an attempt to analyse the trends in costs per hectare of the crops and output prices. The trends in the case of rice, groundnut, and cotton clearly show

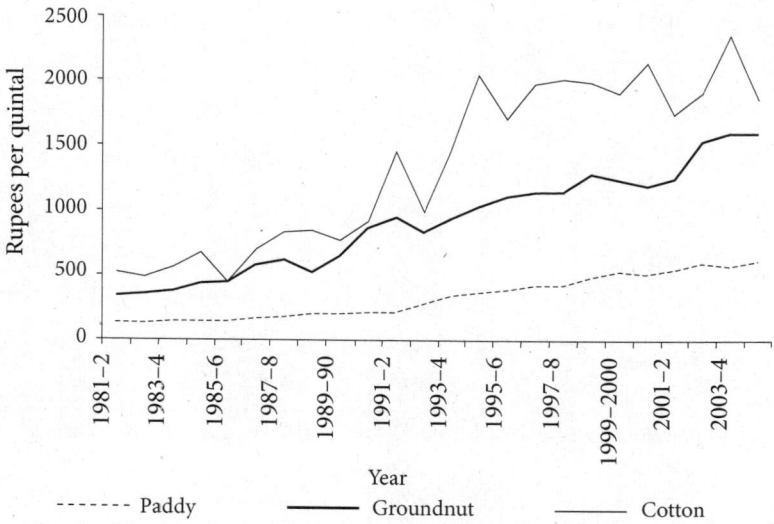

Figure 7.6: Trends in Output Prices in Andhra Pradesh

Source: Cost of cultivation of principal crops in India, Directorate of Economics and Statistics, Department of Agriculture and Cooperation, Ministry of Agriculture, Government of India, New Delhi, various years.

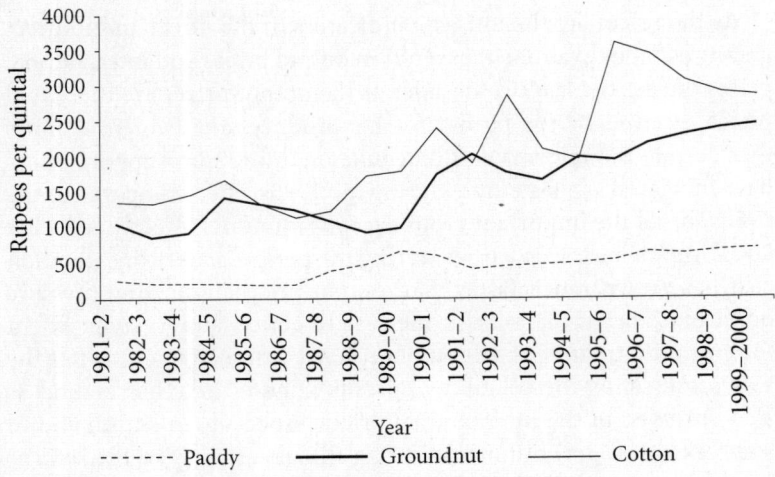

Figure 7.7: Trends in International Prices

Source: Cost of cultivation of principal crops in India, Directorate of Economics and Statistics, Department of Agriculture and Cooperation, Ministry of Agriculture, Government of India, New Delhi, various years.

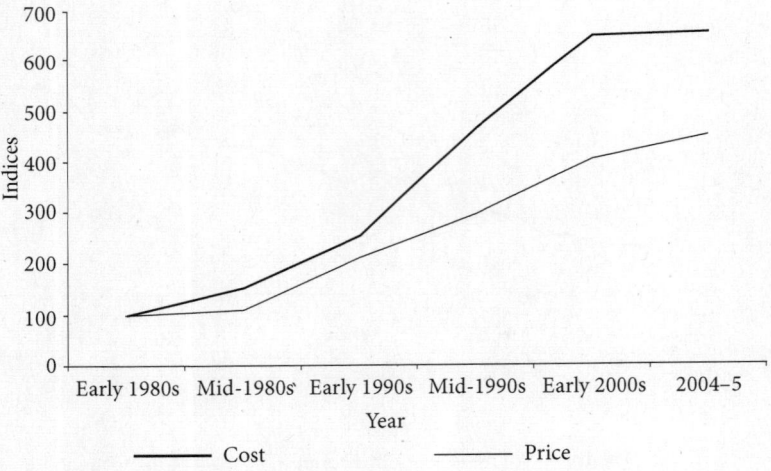

Figure 7.8: Trends in Indices of Costs and Output Prices of Rice

Source: Cost of cultivation of principal crops in India, Directorate of Economics and Statistics, Department of Agriculture and Cooperation, Ministry of Agriculture, Government of India, New Delhi, various years.

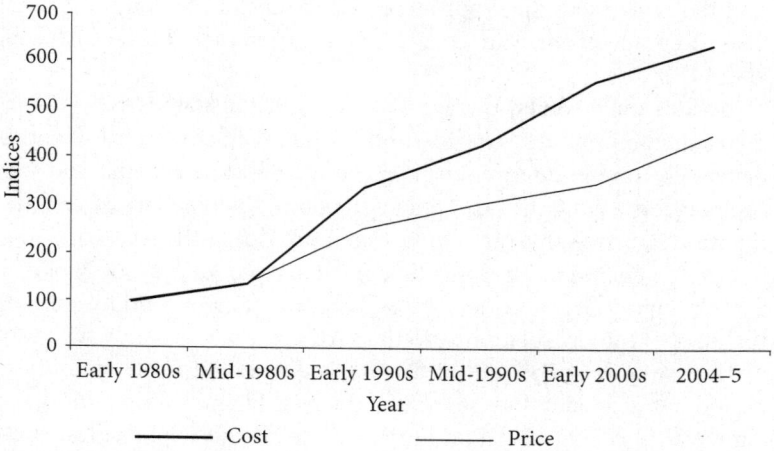

Figure 7.9: Trends in Indices of Costs and Output Prices of Groundnut

Source: Cost of cultivation of principal crops in India, Directorate of Economics and Statistics, Department of Agriculture and Cooperation, Ministry of Agriculture, Government of India, New Delhi, various years.

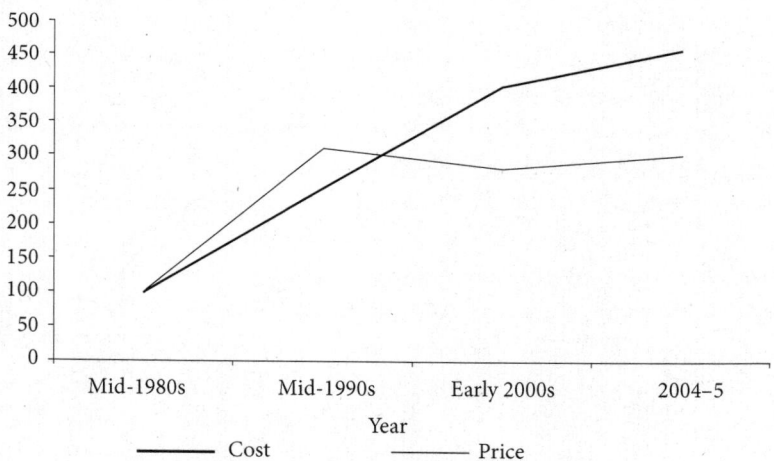

Figure 7.10: Trends in Indices of Costs and Output Prices of Cotton

Source: Cost of cultivation of principal crops in India, Directorate of Economics and Statistics, Department of Agriculture and Cooperation, Ministry of Agriculture, Government of India, New Delhi, various years.

that the costs of all the crops increased faster than the output prices, indicating one of the sources of crisis in agriculture (Figures 7.8, 7.9, and 7.10).

Besides the suicides, the agrarian crisis in the state is reflected in the stagnation or even deterioration in the living conditions of those dependent on agriculture. Even as the overall state income and per capita income rises, the per capita agricultural income faces a decline. Figure 7.11 brings this out clearly. The result is that the levels of living of farmers, including the large holders, languish. Analysis of National Sample Survey Organisation's (NSSO) consumption expenditure shows that the per capita expenditure in the rural areas of the state, which was on the rise between 1983 and 1993, that is, during the pre-liberalization period, came down rather sharply in the post-liberalization period, for almost all agricultural classes in the state. And this happened in spite of an overall increase in the per capita income at the rate of over 3 per cent (Vakulabharanam 2005: 980). The immiserization of the farming community is evident and any small aggravation caused by the pressure from the private moneylender, inability to sell the product because of the depressed prices, or even failure to meet expenses for children's

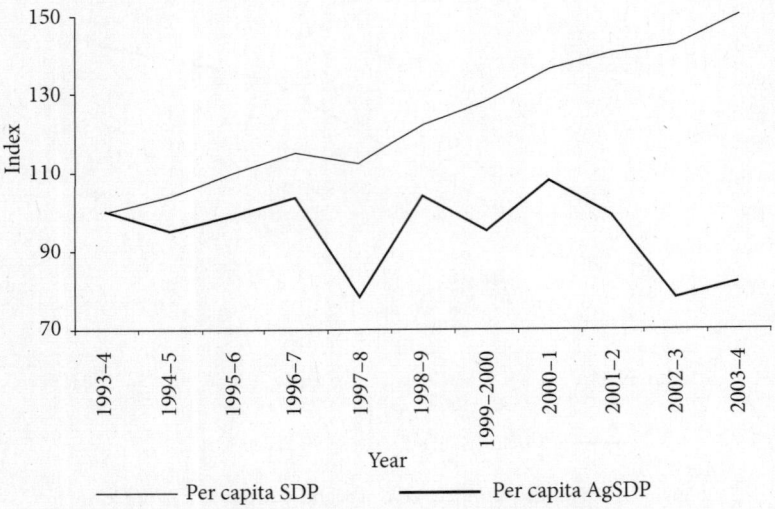

Figure 7.11: Indices of per capita SDP and per capita Agricultural SDP in Andhra Pradesh

Source: Vakulabharanam (2004).

education or meeting health emergencies involving private hospitals or a prolonged drought can trigger the stress beyond endurance, forcing the farmers to commit suicide.

Income and Consumption Levels of Farmers

The precarious situation of farming as a source of livelihood in the country as a whole, and particularly in Andhra Pradesh, is revealed by the farmers' income and consumption levels. Table 7.6 shows that the monthly level of income is less than the monthly consumption expenditure for all classes of farmers up to semi-medium farmers owning four hectares or less, in the state as well as in the country as a whole. This is an unsustainable state for farming to continue as an occupational source. It is no wonder that about 47 per cent of farmers in the state expressed their willingness to quit farming if given a choice. Figure 7.12 shows the levels of average monthly income and consumption for different size-classes of farmers in Andhra Pradesh. Only at the level of medium farmers, owning 4–10 hectares, does the break even between income and consumption expenditure begin, and the level of income rise over and above the consumption expenditure. The situation is similar up to semi-medium farmers for All India as

Table 7.6

Average Monthly Income from All Sources and Consumption Expenditure per Farmer Household: 2002–3

Size-class of land possessed (ha)	Monthly income (Rs)		Monthly consumption (Rs)	
	Andhra Pradesh	All India	Andhra Pradesh	All India
< 0.01	1,107	1,380	2,133	2,297
0.01 – 0.40	1,140	1,663	2,049	2,390
0.41 – 1.00	1,405	1,840	2,274	2,672
1.01 – 2.00	1,837	2,493	2,549	3,148
2.01 – 4.00	2,590	3,589	3,045	3,685
2.01 – 10.00	5,479	5,681	4,133	4,626
> 10.00	9,418	9,667	5,724	6,418

Source: NSSO (2005c).

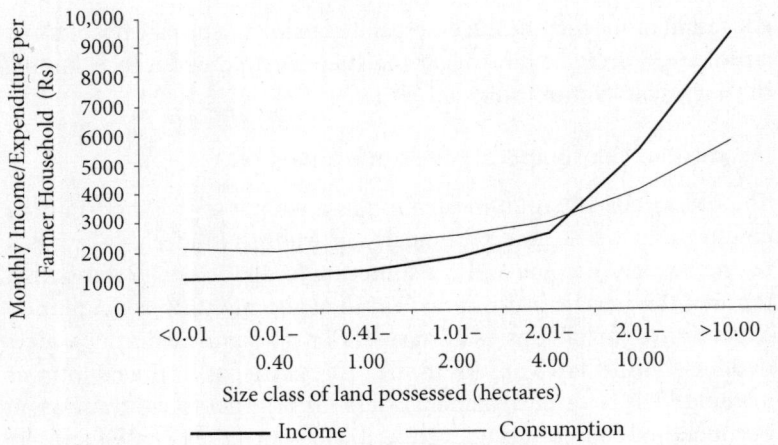

Figure 7.12: Size-class of Landholdings and Average Monthly Income and Consumption Levels per Farmer Household in Andhra Pradesh: 2003.

Source: NSSO (2005b).

well. However, in the case of Andhra Pradesh, both the income and consumption expenditure levels are less than the all-India levels for all classes of farmers.

REFORMS AND RURAL STRESS IN ANDHRA PRADESH

The state of Andhra Pradesh has acquired a special place, in more than one sense, in implementing the economic reforms in all sectors, including agriculture. It was the first state in India to explicitly implement the reform agenda, and it is also the state to reap the largest suicides of farmers in the country. In 1995–6, the then government of Andhra Pradesh ushered in an agenda of economic reforms at the behest of the World Bank's Andhra Pradesh Economic Restructuring Project (APERP). Even the health sector was not spared from the neoliberal economic reforms. Ever since, at least till 2004 when there was a change in political leadership, the state was seen as a kind of laboratory for economic reforms at the provincial level. The economic performance of the state has been subjected to close scrutiny. A considerable amount of analytical contributions on the state of the economy and the conditions of living of the people are available with a focus on the reform period (Rao & Dev 2003). But here the focus is on reforms and their impact in the agricultural sector.

Decline in Public Investment in Agriculture

Agriculture continues to be a major source of livelihood for about two-thirds of populations of the state and still contributes close to one-third of the state's domestic product. Agricultural growth is seen as an important factor in raising rural employment and incomes and in the reduction of poverty. But, during the reform period, agriculture has been in a serious crisis. Gross Fixed Capital Formation (GFCF) in agriculture as a share of total GFCF in the state declined from the peak of 13.83 per cent in 1985–6 to an average of 6–7 per cent in the 1990s. The state experienced a growth rate of 6 per cent per annum in agricultural investment in the 1980s but it decelerated to just 1.5 per cent per annum in the 1990s. There has been steep deceleration in the growth rate of agricultural output from 3.4 per cent in the 1980s to 2.3 per cent in the 1990s. Crop yields in the state have been much lower than in many other states.

There has been a systematic decline in the budgetary expenditure on agriculture and allied activities in the name of fiscal constraint, and alongside a neglect of research and extension. The government investment in agricultural research and education in the state (at 0.26 per cent of its agriculture GSDP during 1992–4) was lower than for the other three southern states and was just around half of that for All India (0.49 per cent for the Centre and states together). Public expenditure on extension, which is borne by the state government, declined in absolute terms in the 1990s. It was only 0.02 per cent of the state's GDP during 1992–4, as against the All-India average of 0.15 per cent. There was an attempt to privatize extension services. As a result of these policies, extension services are currently in bad shape in the state. Agricultural extension services account for only 9 per cent of the farmers' information on agricultural technology in the state. Input dealers (30 per cent) and other progressive farmers (34 per cent) constitute the major source of information (NSSO 2005b).

Extension in Disarray

By the late 1990s, the looming agricultural crisis was recognized to be substantially the consequence of inadequate agricultural services, including extension, reliable seed supply, quality pesticides, machinery, proper soil survey testing, soil conservation, market information, and market intelligence. However, despite this, the Andhra Pradesh state government of that time refused to recognize this and take corrective measures. A 'Working Paper' of the Department of Agriculture (GOAP 1999) stated that the government could act only as a facilitator and no public investment would be made in providing these services. Referring to the vast gap in agricultural extension, because of unfilled vacancies, which at that time accounted for more than one-fourth of the sanctioned posts, it was declared that the state 'doesn't have resources to employ any more extension workers', and so it was proposed that the entire cadre of agricultural extension officers be wound up. 'Without any additional financial burden to the state', the extension services would be promoted through the private sector through a system of registration of unemployed grantees or retired employees, who would offer these services for a fee. Qualified graduates would be encouraged to become licensed dealers of fertilizers, pesticides, and seeds. The burden on the Andhra Pradesh Seed Corporation would be reduced by making the private sector more accountable through an appropriate Memorandum of Understanding (MOU). The hiring of agricultural machinery would be encouraged through the corporate sector, NGOs, and others. Soil surveys, soil conservation, and collection of market information were to be 'encouraged and to be developed in private sector with appropriate policy incentives'.

With the aggressive reformist approach of the state, it is not surprising to find that many public institutions critical for farmers and agricultural development were systematically eroded or destroyed. Some important government corporation and cooperative institutions in the state were closed, allowed to run down, or simply handed over to the private sector. These institutions, such as the Andhra Pradesh Irrigation Development Corporation, Andhra Pradesh Agro-Industries Corporation, Andhra Pradesh Seeds Development Corporation, Cooperative Sugar Factories, and Cooperative Spinning Mills, played an important role in helping the farmers and their closure also affected the farmers adversely.

The failure of extension services, the mushrooming of spurious seed and pesticide companies, and the relegation of the agricultural university and the Andhra Pradesh Seeds Corporation to an insignificant role in the research, development, and propagation of seeds of non-food crops are all the consequences of deliberate policy changes in the state.

Reforms, Agricultural Credit, and Farmers' Indebtedness

The *Situation Assessment Survey of Farmers* of the 59th Round of the NSSO has provided ample statistical evidence on several aspects of farming in India, including 'indebtedness'. The *All India Debt and Investment Survey* 2002 (NSSO 2005d) provides extensive information on farmers' debt. The results of these surveys show that the situation of indebtedness of farmers in Andhra Pradesh is much worse than most of the other states. In 2002, the incidence of indebtedness among farmer households is 82 per cent, which is the highest among all the states. The next highest incidence is 74.50 per cent in Tamil Nadu and the national average of the incidence is only about 48.60 per cent.

One of the most disturbing aspects of farmers' indebtedness in Andhra Pradesh is the unusually high dependence on non-institutional sources. It is well-known that some of the radical banking reforms including nationalization of commercial banks and starting of regional rural banks (RRBs) in the 1970s, vastly improved farmers' access to institutional credit in the 1970s and 1980s. The economic reforms and the reforms in the banking sector since 1991 however, turned the trend towards a decline in the share of institutional credit in the total borrowings of farmers. As a result, the share of institutional sources in the overall debt of rural cultivators in the State was as low as 32.7 per cent in 2002, compared to 67.3 per cent for All India (Figure 7.13). Access to institutional sources for small and marginal farmers is much worse, and there is an inverse relationship between the size of landholding and access to institutional credit (Figure 7.14).

In Andhra Pradesh, the single largest source of debt is 'moneylender' (agriculturist and professional), accounting for 54 per cent of the total debt in 2003 (NSSO 2005a). While cooperatives emerge as the single largest source for 'All India' with about 30 per cent of the loans, in Andhra Pradesh this source accounts for about 15 per cent. The cause for concern with the high share of non-institutional sources is because of the high rates of interest. While most of the institutional loans bear interest rates of less than 15 per cent, a substantial part of non-institutional sources bear interest rates of more than 30 per cent.

Micro-level village studies shows that 92 per cent of non-institutional loans bear interest rates ranging from 24 per cent to more than 60 per cent, and the median rate is 36 per cent (Reddy and Reddy 2007).

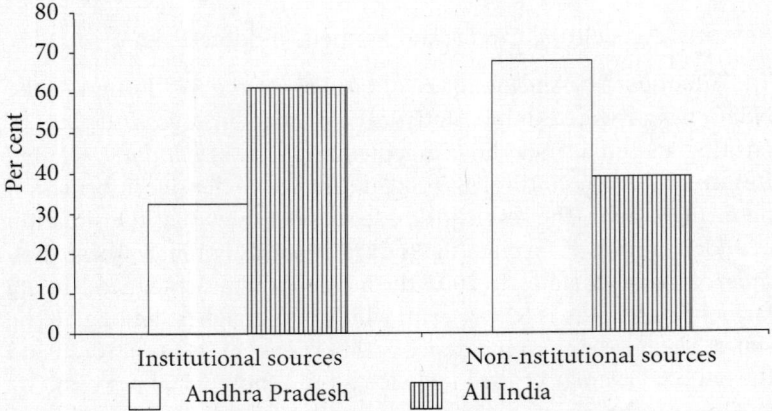

Figure 7.13: Percentage Distribution of Institutional and Non-institutional Loans to Rural Cultivator Households in Andhra Pradesh and All India: 2002

Source: NSSO (2005a)

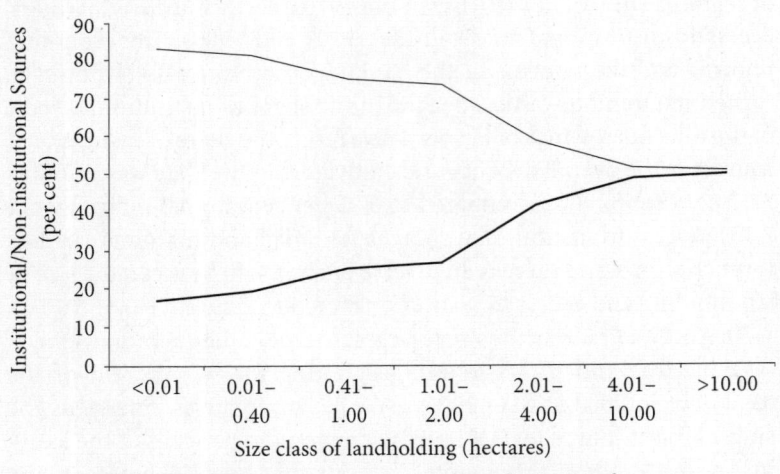

Figure- 7.14: Size-class of Landholding and Institutional and Non-institutional Share of Outstanding Loans: 2003

Source: NSSO (2005a).

An analysis of institutional credit flow in the post-reform period shows that there was decline in rural commercial bank branches in the state from 2644 in 1990 to 2400 in 2004. The share of agriculture sector in the total commercial bank credit declined from 80 per cent in 1993–4 to 50 per cent in 2002–3. The performance of cooperatives and regional rural banks was no better. Of the total institutional credit in 2002–3, cooperatives accounted for a meagre 15 per cent and RRBs 11 per cent. In spite of heavy dependence on high-cost non-institutional sources, the share of total productive expenditure in the outstanding debt of farmers in the state has increased from 46.5 per cent in 1991 to 57.3 per cent in 2002 (Reddy and Reddy 2007).

Concluding Remarks

Given the existing agrarian structure and the resource situation, the important question is whether small–marginal farming is sustainable without substantial public infrastructure support and comprehensive social security including health, education, supplementary employment, and old age support. Even at the early stages of the structural adjustment programmes (SAPs), there was clear warning that neoliberal reforms would force adjustment among poor farmers, which without assistance from the state, would intensify their suffering (Cornia et al. 1987). By and large, the incidence of suicides has been higher among small–marginal farmers moving from subsistence agriculture to the high-value crops with a strong motivation to improve their social and economic status. They are indeed risk-taking small agricultural entrepreneurs whose success would be the basic premise for transformation of the rural India towards better and equitable incomes and livelihood. A keen scholar of the Indian countryside observes: 'farmers' distress is not due to lack of agricultural growth but paradoxically due to enterprising qualities of farmers who pursue growth and even achieve it in good measure. But drought-prone environment and a non-caring policy regime turn those who bring growth into victims' (Rao 2004).

It is a cruel paradox that the state is agriculturally self-sufficient and the policy makers have designs and dreams about high export growth of agricultural commodities, including foodgrains, but farmers who are the architects of these surpluses are allowed to die out of distress. What is needed is a caring policy but what exists instead is exposure to predatory market forces. There is increasing evidence that there cannot be rural development in a predominantly agricultural state such as Andhra Pradesh without high agricultural growth. Nor is there any instance worldwide of dryland farmers moving to high productivity agriculture in the face of gross exposure to the volatile market forces. There is no instance of small–marginal farmers earning adequate livelihood without appropriate social security and economic support nor without succour provided by supplementary non-farm employment. Small–marginal farmers in the dryland regions are the most vulnerable but least cared for in the economic reforms framework. It is the policy neglect that has been forcing these farmers to shoulder all the costs and risks of high investment, including land and water resource development, with borrowed capital at usurious interest rates. They have lifted the state's agriculture to relatively better productivity at a cost that they can ill afford.

These costs are the costs of transition of agriculture in the state from subsistence levels to higher productivity. These costs are necessarily social costs, which should not be compounded on to the shoulders of the distressed peasantry. The state has to own the responsibility for these social costs of investment in the development of land and water (including groundwater) resources, provision of adequate economic support by way of institutional credit, extension, quality input supply, and remunerative prices as well as social sector support of ensuring quality education and health facilities in the countryside. There is incontrovertible evidence that agricultural growth driven by improved productivity of small–marginal farmers would result in a much more equitable distribution of income and augmentation of effective demand with its spread effects on the non-farm sector, and would be more sustainable as well. The essential condition is the need for a policy shift from the neoliberal market-centred reforms to the building of economic and social support systems to make small–marginal farming, especially in dry regions, viable, and to insure these farmers against exposure to distress due to the vagaries of domestic and global market forces.

NOTES

1. The state is divided into three administrative regions, viz., Coastal Andhra, Telangana, and Rayalaseema, based on historical, socio-economic,

and cultural factors. Agro-climatically, the state is demarcated into five zones, viz. North Coastal Andhra (NCA), South Coastal Andhra (SCA), Rayalaseema, North Telangana (NTG), and South Telangana (STG).

REFERENCES

Action for Welfare and Awakening in Rural Environment (AWARE) (1998), *Report on Farmers Suicides in Andhra Pradesh*, Hyderabad: Development Research Advisory Group.

Bhat, K. S. and S. Vijaya Kumar (eds) (2006), *Undeserved Death: A Study on Suicide of Farmers in Andhra Pradesh (2000–2005)*, Hyderabad: Council for Social Development.

Bhushan, S. and T. P. Reddy (2004), 'A Moving into Poverty Syndrome: Debt and Differentiation in Small Farm Economies: A Causal Study of Farmers' Suicides in Andhra Pradesh', mimeo, Hyderabad: Poverty and Social Monitoring Unit (PASMU), Society for Elimination of Rural Poverty (SERP), November.

Centre for Environmental Studies (CES) (1998), *Gathering Agrarian Crisis: Farmers' Suicides in Warangal District (AP): A Citizen's Report*, Hanamkonda (AP): CES.

Cornia, G. A., R. Jolly, and F. Stewart (eds) (1987), *Adjustment with Human Face*, 2 Vols, Clarendon: Oxford, University Press.

Dev, S. M. and C. Ravi (2003), 'Macroeconomic Scene: Performance and Policies', in C. H. H.. Rao and S. M. Dev (eds.), *Andhra Pradesh Development: Economic Reforms and the Challenges Ahead*, Hyderabad: Centre for Economic and Social Studies.

Government of India (various years), *Cost of Cultivation of Principal Crops in India (AP)*, New Delhi: Directorate of Economics and Statistics, Ministry of Agriculture and Cooperation.

Himanshu (2005), 'Rural Income, Employment and Agricultural Growth: An Analysis with Disaggregated Data', unpublished Ph.D. Thesis submitted to the Jawaharlal Nehru University, New Delhi.

Human Rights Forum (HRF), Anantapur (2005), 'Aarugalam Karuvu' (Dry Season Famine), Report on Agricultural Crisis in Anantapur, Hyderabad: HRF, May.

Kumar, S. N. (2005), 'Focus on the Farm Sector', *The Hindu*, 22 December.

Krishna Rao, Y. V. and S. Subrahmanyam (eds) (2002), *Development of Andhra Pradesh (1956–01): A Study of Regional Disparities*, Hyderabad: N.R.R.Research Centre, C.R.R.Foundation.

Mohan Rao, R. M. (2004), *Suicides Among Farmers: A Study of Cotton Growers*, New Delhi: Concept.

National Sample Survey Organisation (NSSO) (2005a), *Situation Assessment Survey of Farmers: Indebtedness of Farmer Households*, NSS 59th Round (January–December 2003), Report No. 498 (59/33/1), New Delhi: Ministry of Statistics and Programme Implementation, Government of India, May.

—— (2005b), *Situation Assessment Survey of Farmers: Access to Modern Technology for Farming*, NSS 59th Round (January–December 2003), Report No. 499 (59/33/2), New Delhi: Ministry of Statistics and Programme Implementation, Government of India, June.

—— (2005c), *Situation Assessment Survey of Farmers: Consumption Expenditure of Farmer Households*, NSS 59th Round (January–December 2003), Report No. 495 (59/33/4), New Delhi: Ministry of Statistics and Programme Implementation, Government of India, October.

—— (2005d), *Household Indebtedness in India as on 30.06.2002: All India Debt and Investment Survey*, NSS 59th Round (January–December 2003), Report No. 501 (59/18.2/2), New Delhi: Ministry of Statistics and Programme Implementation, Government of India, December.

Parthasarathy, G. and Shameem (1997), 'Suicides of Cotton Farmers in Andhra Pradesh: An Exploratory Study', *Economic and Political Weekly*, Vol. 33, No. 13, 28 March, pp. 720–6.

Rao, C. H. H. and S. M. Dev (2003), 'Economic Reforms and Challenges Ahead: An Overview', in C. H. H. Rao and S. M. Dev (eds), *Andhra Pradesh Development: Economic Reforms and Challenges Ahead*, Hyderabad: Centre for Economic and Social Studies, pp. 1–30.

Rao, P. N. and K. C. Suri (2006), 'Dimensions of Agrarian Distress in Andhra Pradesh', *Economic and Political Weekly*, Vol. 41, No. 16, 22 April, pp. 1546–52.

Rao, V. M. (2004), *State of the Indian Farmer: A Millennium Study—Rainfed Agriculture*, Vol. 10, New Delhi: Academic Foundation.

Rao, V. M., and D. V. Gopalappa (2001), 'Agricultural Growth and Farmer Distress: Tentative Perspectives from Karnataka', *Economic and Political Weekly*, Vol. 39, No. 52, 25 December, pp. 5591–8.

Reddy, D. N. (2006a), 'Economic Reforms, Agrarian Crisis and Rural Distress', 4th Annual Professor B. Janardhan Rao Memorial Lecture, Professor B. Janardhan Rao Memorial Foundation, Warangal, Telengana.

—— (2006b), 'Half a Century of Travails of Agriculture in Andhra Pradesh', in R. S. Rao, V. H. Rao, and N. Venugopal (eds), *Fifty Years of Andhra Pradesh 1956–2006*, Hyderabad: Centre for Documentation, Research and Communication.

—— (2007), 'Farmers' Indebtedness and Agricultural Credit Situation in Andhra Pradesh', Background paper for the Expert Group on Agricultural

Indebtedness, mimeo, New Delhi: Banking Department, Ministry of Finance, Government of India.

Reddy, V. R. (2003), 'Irrigation: Development and Reforms', *Economic and Political Weekly*, Vol. 38, Nos 12 and 13, 22 March, pp. 1179–89.

Reddy, V. R. and B. Behera (2003), 'Ecological Divide: Regional Disparities in Land and Water Resource Management in Andhra Pradesh', in C. H. H. Rao and S. M. Dev (eds.), *Andhra Pradesh Development: Economic Reforms and the Challenges Ahead*, Hyderabad: Centre for Economic and Social Studies.

Reddy, V. R. and P. P. Reddy (2003), 'Domestic Trade Policies in the Context of Trade Liberalisation', *Indian Journal of Agricultural Economics*, Vol. 58, No. 3, July–September, pp. 333–52.

—— (2007), 'Increasing Costs in Agriculture: Agrarian Crisis and Rural Labour in India' , *Indian Journal of Labour Economics*, Vol. 50, No. 2, April–June, pp. 273–92.

Reddy, V. R. and S. Galab (2006), 'Farmers' Suicides: Looking Beyond the Debt Trap', *Economic and Political Weekly*, Vol. 41, No. 19, 13 May, pp. 1830–38.

Revathi, E. (1998), 'Farmers Suicides: Missing Issues', *Economic and Political Weekly*, Vol. 33, No. 17, 16 May, p. 1207.

—— (2007), 'Suicide as Vulnerability: Some Dimensions of Agrarian Distress in Andhra Pradesh', mimeo, Hyderabad: Centre for Economic and Social Sciences.

Rythu Sahaya Committee (RSC) (1998), *Farmers Suicides in Andhra Pradesh: Report of the Peoples Tribunal*, Hyderabad: RSC, July.

Sainath, P. (2004), 'Sinking Borewells and Rising Debt', *The Hindu*, 23 June.

Shiva, V. and A. H. Jafri (1998), *Seeds of Suicide: The Ecological and Human Costs of Globalisation in Agriculture*, New Delhi: Research Foundation for Science, Technology and Ecology.

Sridhar, V. (2006), 'Why do Farmers Commit Suicide: The Case of Andhra Pradesh', *Economic and Political Weekly*, Vol. 41, No. 16, 22 April, pp. 1559–65.

Sudhakumari, V. M. (2002), 'Globalization, Economic Reforms and Agrarian Distress in India: a Study of Cotton Farmer's Suicide in Warangal District', Paper presented to the 2002 GDN Award for Research Papers, Thiruvananthapuram: University of Kerala.

Vakulabharanam, V. (2004), 'Immiserizing Growth: Globalization and Agrarian Change in Telangana, South India between 1985 and 2000', Ph.D. Dissertation, Amherst: Economics Department, University of Massachusetts, 2004.

——— (2005), 'Growth and Distress in a South Indian Peasant Economy During the Era of Economic Liberalisation', *Journal of Development Studies*, Vol. 41, No. 6, August, pp. 971–97.

Vidyasagar, R., and K. Suman Chandra (2004), 'Farmers' Suicides in Andhra Pradesh and Karnataka', Hyderabad: National Institute of Rural Development.

World Bank (2003), *Unlocking the Growth Opportunities in Andhra Pradesh*, New Delhi: Poverty Reduction and Economic Management Sector Unit, South Asia Region.

8

Agrarian Transition and Farmers' Distress in Karnataka

R. S. DESHPANDE

INTRODUCTION

Karnataka state has come to acquire an image of being one of the aggressively reforming states of the country. Its advances in the areas of information technology and biotechnology have received accolades the world over, but a lot is hidden behind this scene. Karnataka's agricultural sector has shown signs of retrogression. Though growth *per se* may appear to be apparently delinked from agriculture, the welfare of large masses is not. Karnataka's economy still decisively depends upon the agricultural sector and, as development crisis is located in this sector, the economic growth pattern in the state is virtually dictated by the agricultural sector. Experience has shown that when the agricultural sector of the state fails to perform, sustaining the growth rhythm of the aggregate economy is problematic (Planning Commission 2007). This is mainly due to the vast spread of drought-prone areas with sporadic patches of protective irrigation. Karnataka is second to Rajasthan in its share of drought-prone areas across the states in the country, a fact that is often overlooked. Consequently, a large share of agricultural land of the state is under rainfed conditions with severe agro-climatic and resource constraints. It is one of the least protected agriculture, both in terms of natural endowment and developmental efforts.

A comparison across states, especially the comparable states, reveals some interesting pointers about the performance of agriculture in Karnataka (Annexure 8.1). Karnataka has one of the lowest shares of area under irrigation, underscoring the predominance of rainfed agriculture in the state (Deshpande & Raju 2001). Consequently, the irrigation and cropping intensity in the state is one of the lowest. Except for maize, the productivity of foodgrains is also low. Karnataka has the critical combination with high share of drought-prone area, large proportion of wastelands, high density of marginal and small farmers, and high outstanding agricultural credit from scheduled banks. These conditions have the potential to cause deep agrarian crisis, as the present one with a number of suicides by farmers.

A few questions crop up here. Will the current development strategy sustain the agricultural sector of the state in the liberalized economic scenario? Is it possible for the state to move up in the growth path despite the constraints attached to it? Will the new growth path, with market forces taking the vantage positions, hamper the agrarian transition? Will that require crucial growth drivers to hold ahead a promising situation and attract fresh investment? And finally, what will be the process of agrarian transition in this context?

This chapter attempts to answer some of these questions. The second section analyses the changing contours of agriculture, the slowdown of agricultural growth and decline in investment, particularly public investment in agriculture. The third section draws attention to the increasing regional disparities in agricultural resource endowments and development, and intersectoral imbalances. The fourth section is devoted to the growing marginalization in the agricultural sector. The fifth section analyses the nature and ramifications of farmers' suicides based on a field survey. The last section presents policy alternatives to move out of the crisis.

OVERVIEW OF AGRICULTURE: CONTOURS OF CHANGE AND CHANGING CONTOURS

The agrarian structure in Karnataka has undergone significant transformation during the last five decades. These changes can be categorized into two broad groups—first, those influencing the macro-economic performance of the agricultural sector largely governed by the policy initiatives and, second, those caused by the institutional changes that have taken place during last three eventful decades. Looking at the macro-economic changes, it can be seen that

Karnataka has changed faster in its macro-economic parameters compared with similar states in the country (Deshpande et al. 2004). The fiscal discipline was tightened with the White Paper on the State Finances (2000) and the Medium Term Fiscal Plan was put in place (Government of Karnataka [GoK] 2006). This helped to bring significant control on public expenditure and kept the fiscal deficit in check. While budgets were balanced and surpluses shown, in the process, however, public expenditure on agricultural sector declined. This resulted in inter-sectoral growth imbalances, and consequently, the macro performance of the agricultural sector was affected. On the one hand, positive changes in the macro economy in the state occurred rapidly but, on the other hand, these changes were accompanied by significant negative externalities inflicting net welfare loss in the agricultural sector (World Bank 2004). Second, institutional changes introduced during the 1970s had created a conducive stage for growth of the state's agricultural sector . These initiatives sustained the sector for long but failure to read the undercurrents of the changes in the state economy ended up in adverse consequences for the state by the end of the 1990s. The development model of Karnataka is undergoing brisk changes, and the mood is to further squeeze investment in the agricultural sector (Government of Karnataka 2007b).

Agricultural Growth

Agriculture and allied activities constitute 17 per cent of the Gross State Domestic Product (GSDP) of Karnataka, and 69 per cent of the workforce of the state is engaged in this sector. Within the sector, crop husbandry accounts for 32 per cent of the Net State Domestic Product (NSDP) originating from agriculture. The share of agricultural sector in GSDP declined by half in the 1990s—from 33 per cent in 1993–4, to 17.5 per cent in 2003–4. There is only a marginal shift of workers out of agricultural sector. Consequently, the GSDP generated per worker in the sector is declining in real terms. The growth performance of the state during last four decades shows fluctuations across sectors. GSDP of the state grew at 5.4 per cent per annum during 1980–1 to 1989–90. The growth rate improved to 7.3 per cent per annum during 1993–4 to 2003–4. The growth rate of GSDP originating from the agricultural sector, however, declined from 2.7 per cent per annum in the 1980s to 1.9 per cent during 1993–4 and 2003–4.

Table 8.1

Growth Performance of State Domestic Product

Sector	GSDP		NSDP	
	1980–1 to 1989–90	1993–4 to 2003–4	1980–1 to 1989–90	1993–4 to 2003–4
Primary	2.7	1.9	2.6	1.9
Secondary	6.7	8.1	6.7	7.4
Tertiary	7.5	10.2	7.6	10.7
All	5.4	7.3	5.2	7.1

Note: GSDP denotes Gross State Domestic Product and NSDP denotes Net State Domestic Product.
Source: Calculated from data obtained from the Directorate of Economics and Statistics, Government of Karnataka, Bangalore.

Capital Formation and Budgetary Support

During the period 1993–4 and 2003–4, the fastest growth was seen in the tertiary sector, followed by the manufacturing sector. The growth

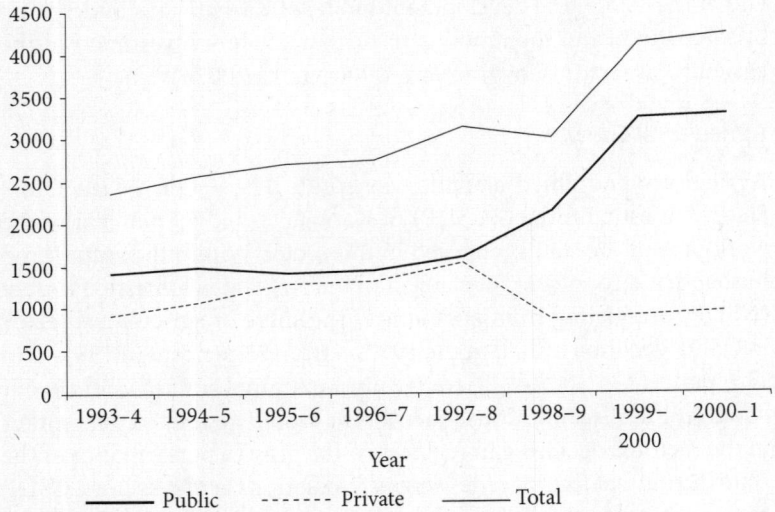

Figure 8.1: Estimates of Gross Fixed Capital Formation by Sources
in Karnataka Agriculture

Source: Agricultural Statistics at a Glance, Ministry of Agriculture, Government of India, various years.

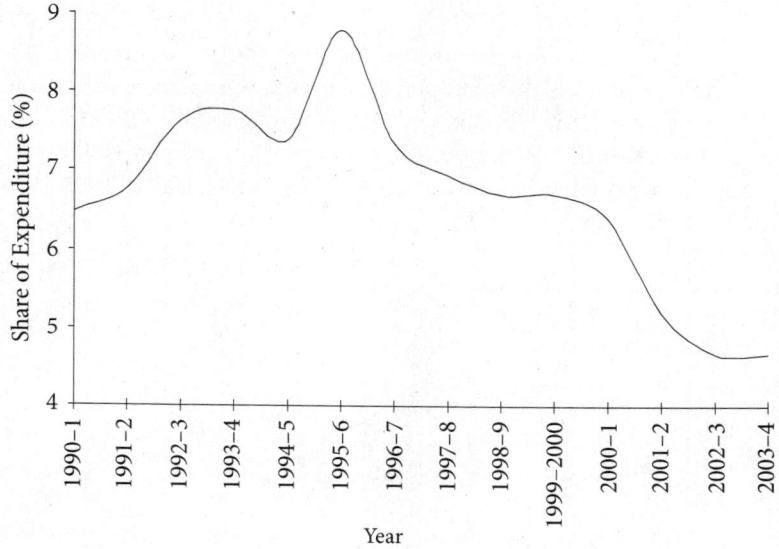

Figure 8.2: Share of Budgetary Expenditure on Agricultural Sector to
Total Expenditure

Source: Government of Karnataka (2005b) and Government of Karnataka (various years)

of the agricultural sector, however, was very low during this decade. There are four broad reasons for the poor performance of agriculture. First, capital formation in agriculture has been more or less stagnant during the mid-1990s, 1993–4 to 1998–9 (Figure 8.1). Second, the decline in the share of budgetary support to the sector also played a significant role in staggering the process of growth (Figure 8.2). There is a strong complementarity between public and private investment in agriculture. Public investment as a stimulus to promote private investment was lagging in this period. Third, the performance of agriculture in terms of its share in the SDP depends not only on production but also on agricultural commodity prices, which were depressed during the period. Lastly, there are regional differences in the agricultural development of the state. 'High growth pockets' need revival in agricultural strategies, whereas the 'lagging regions' need substantial investment in infrastructure, and water resource conservation and augmentation.

Farm Incomes

A comparison of the trends in the real per capita income for agriculture sector (RPCIAS) generated from agriculture sector and the Consumer Price Index for Agriculture Labour (CPIAL), also indicates a disturbing picture. The per capita income generated from agriculture remained almost stagnant during the decade 1992–3 to 2002–3 (Figure 8.3).

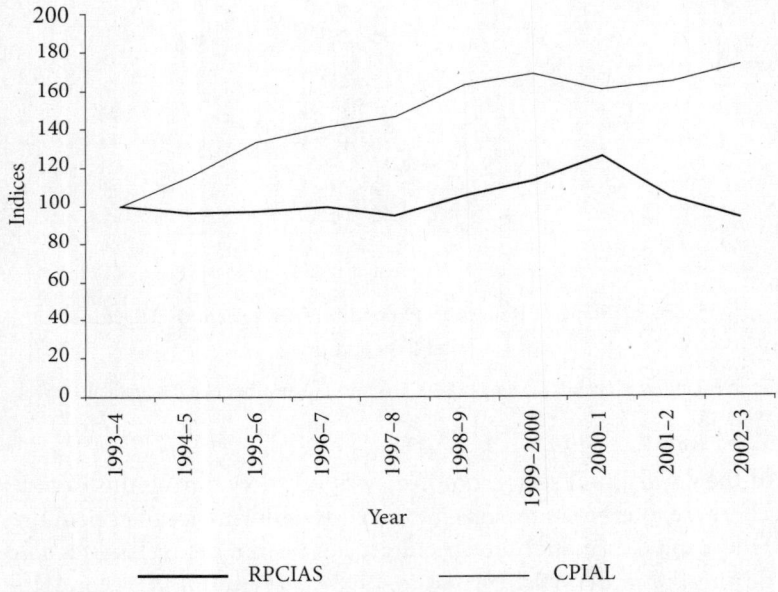

Figure 8.3: Movements in the Index of RPCIAS and CPIAL in Karnataka:
1993–4 to 2003–4

Note: RPCIAL: Index of Real Per Capita Income of Agricultural Sector (includes Agricultural Labourers and Cultivators) (at 1993–94 prices); CPIAL: Consumer Price Index for Agricultural labourers (base: 1993–94= 100).
Source: Deshpande (2005).

From Figure 8.3, it clear that the index for RPCIAS remained largely stagnant, whereas the CPIAL has recorded an increase. This provides evidence for the slowing down of income generation and increasing distress in the agricultural sector. The high volatility in the wholesale prices with an underlying trend of depressed state of the year-to-year

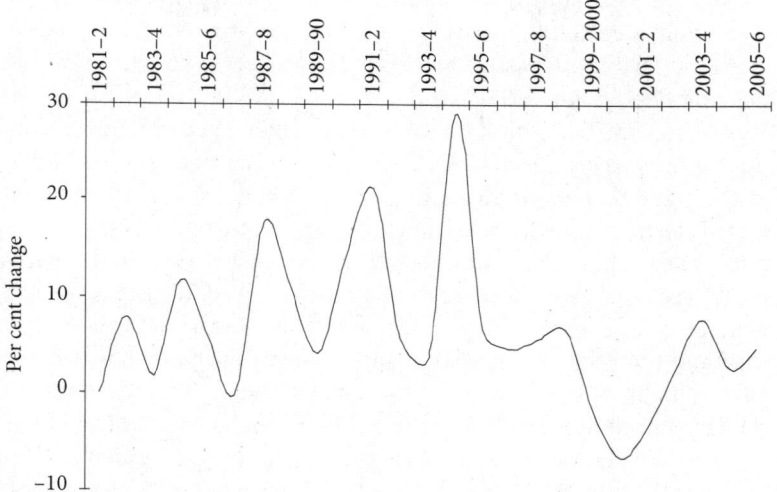

Figure 8.4: Annual per cent Change in Wholesale Price Indices of
Agricultural Commodities

Source: Government of Karnataka (2007a).

movements and consequent shrinking of the flow of income to the
farmer are also evident (Figure 8.4).

Among the macro policy initiatives suggested in the Agricultural
Policy 2006, (Government of Karnataka 2006) additional investment
in the agricultural sector stands out as a critical policy input. Further,
the investment needs to be directed towards the backward regions
identified by the High Power Committee for Redressal of Regional
Imbalances (HPCRRI) (Government of Karnataka 2002). These regions
are showing signs of stepping into new avenues of growth but lack the
basic infrastructure and access to market. Reforms in the domestic
market are also of importance here.

Green Revolution and Narrow Techno-Centred Development

The crisis created by the successive droughts and food insecurity
of the mid-1960s ushered in the Green Revolution, which brought
significant growth in the agricultural sector. As in other states,
Karnataka too recorded an impressive performance in the agricultural
sector due to the Green Revolution (Nadkarni & Deshpande 1983,

p. 30). It is well known that the Green Revolution had essentially a technology-centred approach focussing on high-yielding seeds, chemical fertilizers, pesticides, and technical management. While in its early days, doubts were expressed about the impact of the Green Revolution, the fact is that it provided the much needed food security during those days (Deshpande et al. 2004). It has been observed that in the green revolution phase, the agricultural development has been largely techno–centric, relegating the farmer to the backstage as a mute participant. There has been a mismatch between investments in technology generation and the farmers' welfare. Many of the generated technologies have not reached the farmer effectively, and in many there has been a large gap between the potential and the achievements (Government of Karnataka 2005a). The ambitions of the farmers were propelled and when these could not be met, a huge gap was created between the expectations and the experience. This was largely due to the techno-centric approach concentrating on inputs and productivity rather than income generation. The failure of this techno-centric approach to enhance the income flow became fully exposed due to market failure. Consequently, the net income from the farm sector has stagnated resulting in the current phase of farmers' distress. Therefore, over-reliance on the techno-centric approach has some important consequences in the context of the recent agrarian transition.

In view of the slowdown in yield growth (sometimes negative) during the 1990s across the country, the total factor productivity growth in Karnataka has also decelerated (Planning Commission 2007) (Annexure 8.2). The exclusive dependence on seed–fertilizer approach to productivity growth, with inadequate emphasis on organic treatments, has led to depletion of soil fertility and inadequate environmental support for a sound farming system. The emphasis has been more on technology than the backward and forward linkages for agricultural development. The approach also neglected the environmental support for establishing a sustainable farming system.

The explanation of this stagnation in productivity through the climatic aberrations during the last three years of the 1990s is quite obvious and not to be contested. But this delicate sensitivity to the slight climatic aberrations bares the continued vulnerability of the state agriculture and derides the decades of research on dry farming practices as well as failure of technology to provide drought tolerant varieties and technology. The stagnation has been rather perpetual in the case of oilseeds and pulses, which received least attention in the research agenda of the agricultural scientists as compared to cereals, as seen in studies conducted by the two state agricultural universities (Government of Karnataka 2005a). Research in the state agricultural universities largely concentrated on cereals and horticultural crops, followed by the cash crops, while pulses and oilseeds received the least attention (Government of Karnataka 2005a). These two crop groups are largely grown by small and marginal farmers, besides being important crops of Karnataka in the cropping pattern.

Technological innovations are necessary for the farm sector to move ahead in the process of development. The innovations, however, are not simply an end in themselves, but constitute only one component of the structured process of innovation that leads to development. This has to be followed by four stages. First, the innovation has to be demand generated and acceptable to the farmers in the current cultural and resource availability context. Those innovations that cross the cultural boundaries and have unaffordable resource requirements mostly stay in the laboratory itself. Second, the new technology has to reach the farmers' field without time lapse. Delay in this regard may mean denial of the technology to a large number of potential users. The phrase 'lab to land', has been repeatedly used, nonetheless little seems to have happened on this count. The technologies developed in the agricultural universities take a longer time to reach the farmer as compared to those developed by the private agencies almost in similar conditions (Government of Karnataka 2005a). This is a serious lapse. Third, the task of the research institutions in the public domain does not end merely by introducing a new crop or a variety; it is also necessary that the ultimate consumer accepts it, something that is always left unattended. Besides this, the push effect (pushing the traditional crops out of cultivation) that these new crops/varieties exert is seldom analysed. In the pursuit of the new cultivars even the better quality local crops are neglected, so also the intrinsic traits of these local breeds are forgotten. In the push effect, what is lost is not only the precious traditional germ plasma but also the time-tested mechanism to deal with climatic aberrations. Fourth, the produce of the new technology has to be taken to the market, and out of every innovation the farmer expects to improve the net returns. But as the traditional adage goes, the professional efficiency of the farmer takes a hard beating at the market; higher arrivals at the market depress the prices of the produce. As a result, the net income of the farmer is reduced and this is at the cost of huge investment in the development of the technology at the agricultural universities as well as its adoption by the farmer (Government of Karnataka 2005a). Paradoxically, even as the investment in agricultural research increases, the farmers' incomes take a beating. Last, the cost of cultivation (especially the cash component) of the new technology is also quite high and that is not absorbed by equivalent increment in the prices received. Moreover, the new cultivars require higher labour input and the wage bill increases.

The point made is not against technology but to emphasize that a techno-centric approach alone cannot be a solution for the problems. In fact, a blind techno-centric approach may fuel the agrarian crisis further, which seems to have happened during the last decade. During the mid-1960s there was a need for a quick solution, as the food insecurity was acute. Therefore, increasing productivity of food crops and especially cereals was concentrated upon. However, this was accompanied by a strong wave of commercialization and increasing cost of cultivation, thereby shrinking net returns in farming (Nadkarni 1988). Therefore, the technology development needs to be viewed pragmatically and a paradigm shift is needed in the agricultural sector, especially concentrating on the bypassed regions and crops.

REGIONAL AND SECTORAL DISPARITIES

Regional disparities within a state often appear to be one of the factors behind farmers' distress. The experience across the country wherever the farmers' distress had been acute, viz. Vidarbha in Maharashtra, Sangrur-Mansa region of Punjab, Rayalaseema and parts of Telangana in Andhra Pradesh and northern Karnataka, shows historical neglect in development to be common among them. These regions were pushed back on the developmental scale. Karnataka is a state formed by merging portions of erstwhile Hyderabad State, Bombay Presidency, Madras state, and old Mysore. These regions had historically demonstrated differential response to developmental initiatives, and hence at the time of merger, they were at varied planes in the development ladder. Efforts were made to bring a regional concordance in their development and growth but the success of these initiatives has been limited more due to the design of the schemes than the efforts *per se.* Therefore, regional disparities were not only sustained but also fuelled further.

The distress among farmers is acute in northern Karnataka. The main reasons for concentration of distress in this region are historically inherited backwardness, relatively poor resource endowment with high drought proneness, and the relative neglect of the region in the developmental efforts. This region is so critically on the margin of land and water resources that even a slight weather disturbance pushes farmers into distress. The Government of Karnataka, over the past four decades, appointed four committees to examine imbalances in development across the state's regions and identify such regions (Deshpande & Dadibhavi 2005). The first such effort was undertaken in 1954, by setting up a Fact Finding Committee to make an assessment of the level of development in various regions of the state (Aziz & Krishna 1996). The Committee proposed to set up statutory Boards for four regions, but the state failed to implement this decision. In 1980, the Hyderabad–Karnataka Development Committee was set up and it suggested a strategy for the development of the erstwhile Hyderabad–Karnataka region. Another committee, consisting of all Secretaries headed by the Chief Secretary, was set up to examine the

Table 8.2

Identification of Developed and Backward Talukas Based on Comprehensive Composite Development Indices (CCDI) in Karnataka: 1975–6 and 2000–1

Regions	No. of talukas	1975–6		2001	
		Developed	Backward	Developed	Backward
North Karnataka	81	31	50(16)	22	59(26)
South Karnataka	94	41	53(5)	39	55(13)
Karnataka state	175	72	103(21)	61	114(39)

Note: CCDI for 1975–6 is based on 17 variables; CCDI for 2001 is based on 35 variables. Figures in parenthesis indicate the number of highly backward talukas.
Source: Dadibhavi (1982), Government of Karnataka (2002).

discrimination across regions in the state during the late 1990s. However, nothing tangible came out of the Report of the Committee submitted in 1999. The efforts of these Committees are well documented in Aziz (1996). The fourth one is the High Power Committee for Redressal of Regional Imbalances (HPCRRI) (Government of Karnataka 2002). Comparison of the HPCRRI identified backward talukas in 2001 with that of an earlier study of 1975–6 (Dadibhavi 1982) is quite revealing. Table 8.2 shows that in 1975–6 there were 50 backward talukas in northern Karnataka and 53 backward talukas in southern Karnataka. This number increased to 59 and 55, respectively, by 2001. In terms of the proportion of backward talukas also, the per cent of backward talukas identified in north Karnataka increased more than those in south Karnataka.

A comparison of the levels of living across the regions in Karnataka, leads to the conclusion that the disparities in development have a telling effect on the differential levels of living. The avenues of income are few and as a result, the consumption expenditure is low. The HPCRRI (2002) identified 114 *talukas* at different levels of backwardness. The list has increased in size over the years and the talukas that were at the borderline during the earlier years have slid down, revealing the fact that there has been an increase in disparities instead of a reduction. These regions stayed backward probably due to the faulty design of the policies. It is not just coincidence, therefore, that the distress in

Table 8.3

Levels of Consumption and Rural Poverty across Regions of
Karnataka, 1999–2000

Region	Average per capita consumption (Rs/month)	Average calorie intake (Kilo calories per day)	Percentage of poor population
Coastal	645.39	2195.3	3.85
Inland Eastern	595.26	2066.2	3.35
Inland Southern	545.34	2046.8	10.73
Inland Northern	431.12	1984.3	25.14
Total	499.44	2028.8	16.90

Source: Government of Karnataka (2001b).

the farm sector is quite acute in the talukas of northern Karnataka, where there has been relative neglect of not only agriculture but other sectors as well, resulting in greater intersectoral imbalances as well.

Sectoral Imbalances

The aggregate nature of regional imbalances, at times, mask the intersectoral divergences and intrasectoral disparities. In some of the regions of Karnataka, we find that some sectors are developing very well as against others. There are resource- as well as policy-induced impediments in the development of these sectors, especially in backward regions. In the state, while the tertiary sector has been growing at a very fast rate, the pace has been varied across the regions. Similarly, the secondary sector has grown only in a few regions, creating a few pockets of prosperity. Transfer of labour, as anticipated in a developing economy, does not function because the secondary sector development itself is quite meagre and the transfer of labour from agriculture to the non-agricultural sector within backward regions is still largely non-existent. Transfer of labour takes place in terms of rural to urban migration, and largely to the cities of Bangalore, Mumbai, and Pune. The phenomenon is quite strong in the northern districts of Bidar, Bijapur, Chitradurga, and Raichur districts. The growth rates in the district domestic product along with the sectoral growth in the regions at two points of time namely, 1980–1 and 1992–3 (at 1980–1 prices) and 1993–4 to 2003–4 (at 1993–4 prices) shows growth in north Karnataka to be slower than that of the state as a whole (Figure 8.5). Across regions, the disparity in growth has been quite high, especially in the growth rates in the services sector and the secondary sector (Deshpande & Dadibhavi 2005).

It is clear that policy emphasis has to be placed on these sectors in order to provide investment incentive in these regions. The average per capita income for the north Karnataka region is continuously moving downward from the state average over the last four decades. The relative figure in terms of district per capita income in all the districts of north Karnataka has moved downward since the 1980s. Not a single district in this region had income above the state average in 2000–1 (Deshpande & Dadibhavi 2005).

The natural resource constraints cannot be the only explanation of the farm sector distress in north Karnataka. It began with

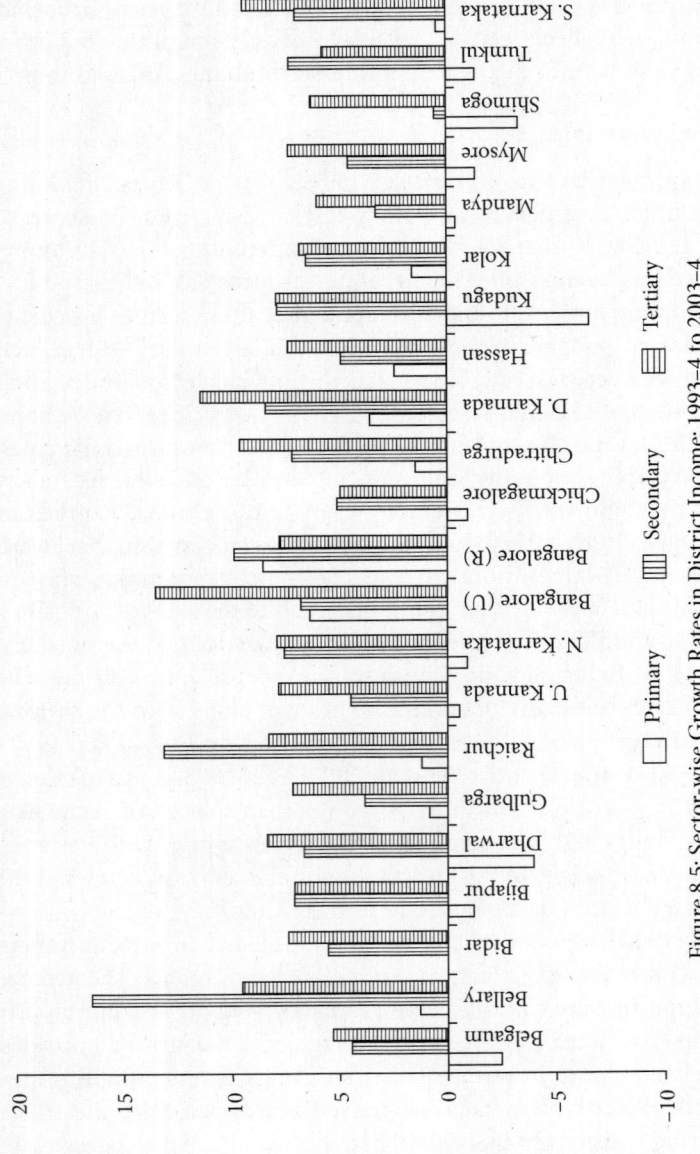

Figure 8.5: Sector-wise Growth Rates in District Income: 1993–4 to 2003–4

Source: Government of Karnataka, Directorate of Economics and Statistics, Bangalore.

historical emergence of the agrarian structure in these regions and was aggravated further by the policy neglect. The region is at such a critical juncture that the technological innovations mostly suitable for better-endowed regions fail to perform in this region. The neglect of infrastructure does not attract investments into other sectors as well and sets in motion mutual reinforcement of a process of marginalization in the region.

AGRARIAN STRUCTURE AND MARGINALIZATION

The agrarian structure of Karnataka is characterized by a large share of small and marginal farmers. Land reforms during the 1970s were quite successful in establishing a new class of peasants but soon the demographic pressure coupled with changes in the land market led to the course of marginalization. The process of marginalization begins with the skewed asset holding, leading to poor quality of assets, and finally results in impacting net income generation. Like elsewhere in the country, Karnataka has also been facing the problem of an increase in the number of marginal and small holdings albeit a little differently (Rao & Hanumappa 1999). That sets the stage for agrarian crises. The process of this crisis begins with five factors. First, the demographic changes initiate a division of land and the land owned gets sub-divided over generations. Consequently, the average size of holding goes down. Second, the non-viability of farming as a profession pushes a large number of farmers to sell portions of the land, and thus their holdings become smaller. Third, with the increasing number of small and marginal farmers, the marketable surplus is generated in small quantities, and the market is discriminatory against the small producers. Fourth, the development of technology also favours the medium and large holdings, as also the market. That brings down the flow of income to these vulnerable groups. Last, the ambitions of the non-viable marginal and small farmers are fuelled by the demonstration effect of other regions and that pushes them to less known technologies (Rao 1991). That makes them to invest in technologies, failure of which creates the crisis.

Land reforms in Karnataka are termed as radical reforms when compared with many other states in the country (World Bank 2000). The reforms were certainly pragmatic in their content but the process of implementation left a large area unattended to and probably fuelled the process of marginalization in the recent past. The major achievements included the acquisition of surplus land, abolition of intermediaries, and abolition of tenancy. The main failures listed by analysts in the context of reforms are the distribution of surplus land, quality of the surplus land, economic viability of the distributed land, and bringing a check on the concealed tenancy. The reform measure such as the consolidation of holding was not implemented in the state. A significant change has taken place in the agrarian situation during the last five decades, as can be observed from the data thrown by the successive agricultural censuses.

While there has been a decrease in the medium and large sized holdings, there has been a large increase in the small and marginal

Table 8.4
Distribution of Landholdings by Size-class in Karnataka:
1970–1 and 2000–1

Particulars	< 1 ha	1–2 ha	2–4 ha	4–10 ha	>10 ha	All
Per cent of number of holdings						
1970–1	30.44	23.65	22.19	17.54	6.16	100
1980–1	34.55	24.53	21.30	15.36	4.24	100
1990–1	39.16	27.45	20.13	11.01	2.23	100
2000–1	45.93	26.96	17.78	8.03	1.27	100
Change per annum (%)	6.69	4.24	1.99	−0.29	−1.96	3.31
Per cent of area operated						
1970–1	4.82	10.74	19.39	33.35	31.67	100
1980–1	6.24	13.13	21.89	34.20	24.51	100
1990–1	8.70	18.73	25.97	30.59	15.99	100
2000–1	12.12	22.28	27.86	26.95	10.78	100
Change per annum (%)	5.73	4.15	1.85	−0.42	−2.1	0.28
Average operated area in hectares						
1970–1	0.51	1.46	2.80	6.09	16.43	3.20
1980–1	0.49	1.46	2.80	6.07	15.69	2.73
1990–1	0.47	1.46	2.75	5.93	15.28	2.13
2000–1	0.46	1.44	2.72	5.83	14.74	1.74
Change per annum (%)	−0.33	−0.05	−0.1	−0.14	−0.34	−1.52

Source: Government of Karnataka (2001a).

holdings (Table 8.4). The small and marginal holdings increased from about 64 per cent in 1970–1 to about 73 per cent by 2000–1 with the growth rates of 4.29 to 6.69 per cent per annum. That indicates the rate at which the agrarian economy is moving towards non-viability. It is not only that the small and marginal holders lack viability but also that the technology generation also failed them. Interestingly, the process of marginalization has been quite sharp during the decade of the 1990s and among the Scheduled Caste (SC) and Scheduled Tribe (ST) farmers (Thorat & Deshpande 1999).

FARMERS' SUICIDES: THE PHASE OF ANGUISH

The distressing phase of agrarian crisis in Karnataka began in 1997 and reached its peak by 2002. It resulted in the spate of suicides by the farmers. We analyse here the context, nature, and causes of suicides based on a small sample of 105 households of farmers who committed suicides across 18 districts of the state (Table 8.5).[1] The study conducted for the Veeresh Committee Report at the Institute for Social and Economic Change (ISEC) (Deshpande 2002a) examined the crisis of farmers' suicides equipped with the micro-level data

Table 8.5

Regional Spread of Suicides across Districts in Karnataka: 2000–1

District	Suicide cases reported	District	Suicide cases reported
Mysore	2	Belgaum	5
Bidar	2	Gulbarga	5
Chikmagalur	6	Bijapur	2
Tumkur	2	Bellary	7
Hassan	8	Koppala	2
Kolar	8	Dharwad	9
Uttara Kannada	4	Haveri	16
Davangere	10	Gadag	4
Chitradurga	9	Shimoga	4
South Karnataka	55	North Karnataka	50
Total 105			

Source: Deshpande (2002a).

collected personally from the victims' families. The distribution of suicides indicates larger density per district in north Karnataka and highest incidence in Haveri district. Davangere, Hassan, and Kolar have also shown higher incidence.

Before we get into the causes of distress, it is essential to highlight a few changes that have occurred during the five decades after independence. First of all, the village as an institution has crumbled down under the pressure of commercialization, whereby the 'weak' in the villages are left to fend for themselves, and the village institutions that hitherto took care of those under distress have slowly receded and vanished. The pressure of commercialization has not only fuelled the weakening process of village institutions but also compartmentalized the classes and even impacted the technology or information transfer. Second, the process of land reforms has created more distortions than solutions. The trends in marginalization of landholding are, as already pointed out, disturbing and, combined with the declining public investment in agriculture, glaring disparities in rural infrastructure and state support to agriculture, exposed an average farmer to unviable risks and uncertainties. It is an irony that the viability threshold has been on the increase and reached about 7 acres while the average farm size has been declining and reached 8.88 acres by 2001 (University of Agricultural Sciences 2001). The demographic pressures have added to this process. More importantly, when the new systems and fresh initiatives were created, enough care was not taken to provide for protection of vulnerable groups and resources. Extension, credit, input delivery system, input markets, product market mechanisms, and the government support schemes, including safety nets, are easily amenable to failure if proper institutional arrangements are not made. The inadequacies are sufficiently pointed out by various expert bodies of the Karnataka Government (Joint House Committee, Agricultural Commission, Agriculture Policy, Chairman: T. R. Satish Chandran, 1995), but not attended to at the implementation level.

Some of the important causes for crisis in agriculture resulting in farmers' distress identified by various expert bodies, in summary, are the following: (i) farmers of Karnataka frequently confronted distress conditions in the form of drought and failure of rainfall. These conditions are severely felt in north Karnataka; (ii) loss of crops due to inferior quality of inputs and non-availability of these inputs on time have been a recent phenomenon. Moreover, adoption of new technology with inadequate knowledge, expertise, and state support created a situation of uncertainty; (iii) sudden attack of pests and diseases and associated economic costs for meeting this exigency disturbs the cost calculus of farmers and pulls down the net expected returns; (iv) yield or productivity loss due to the above reasons or any other reason batters the farmers' expectations about the income flow and leaves them under financial stress; (v) volatility in market prices, non-availability of proper marketing infrastructure, and imperfections in the markets affect the returns from agriculture and income flow to farmers; (vi) mounting credit burden, debt trap, and consequent financial non-viability and interlocked input–credit–product markets have added to the distress of farmers; (vii) failure of extension services in recent years have added to the risks of farmers; and (viii) counselling failure by the institutions and breakdown of traditional village institutions have contributed to increasing stress on farmers, taking the form of suicides in extreme cases.

Borrowing, Credit, and Repayment

Though indebtedness is known as one of the major reasons for farmers' distress, it is the growing dependence on informal sources of credit that charge very high rates of interest and a declining share of formal sources that often puts severe stress on farmers. Table 8.6 shows that dependence on non-institutional sources in households experiencing suicides, especially in the case of marginal, small, and medium farmer households, is much more than in the control group households. Further, the amount borrowed by the victims' families was higher as compared to the control group. The interest rates charged by non-institutions such as moneylenders in some cases were as high as 48–60 per cent. The farmers under distress are driven to non-institutional sources partly because of the inadequacy of the credit extended by the institutional sources and partly because of procedural and other transaction costs involved in institutional credit. Table 8.7 shows that in the case of suicide households, the tendency to borrow from non-institutional sources appears to increase with the increase in the amount of credit needed. In the case of outstanding debt of less than Rs 25,000, 67.5 per cent is from institutional sources, whereas 70.84 per cent of the outstanding debt of Rs 1 lakh and above is from non-institutional sources.

Table 8.6

Distribution of Average Outstanding Debt for Suicide and Non-Suicide
Farmer Households by Size-class across Source

(in Rs)

Size classes	Control households		Suicide households	
	Institutional	Non-institutional	Institutional	Non-institutional
Up to 1 acre	14,167	10,500	5,125	13,750
	(57.43)	(42.57)	(37.27)	(62.73)
1 to 2 acres	22,643	29,095	49,587	65,522
	(43.76)	(52.24)	(43.08)	(56.92)
2–4 acres	20,286	27,893	25,948	52,103
	(42.11)	(57.89)	(33.24)	(66.76)
4–10 acres	31,750	30,321	35,851	49,179
	(51.15)	(48.85)	(42.16)	(57.84)
Above 10 acres	575,000	34,688	46,545	129,364
	(16.58)	(83.42)	(26.46)	(73.54)
All	113,308	28,879	34,847	59,879
	(79.69)	(20.31)	(36.72)	(63.21)

Note: Institutional sources include commercial banks, regional rural banks, and cooperative institutions. Non-institutional sources include moneylenders, relatives, dealers, etc. The cases where the respondents could not distinguish between institutional and non-institutional sources are not included here.

Figures in parentheses indicate percentage distribution between institutional and non-institutional sources.

Source: Deshpande (2002b).

Among the institutions from which the farmers borrow, the Regional Rural Banks (RRBs) are more preferred as against the commercial banks or co-operative banks (Deshpande 2002b). More often, farmers borrow from multiple sources and as a result end up being heavily indebted. It will be beneficial if the credit is recorded in the Credit Record Pass Book permanently maintained by the farmer for the same purpose or entered into the existing Ration Card on a page specifically designated for entering credit availed. Simplification of the disbursement rules and the application procedure by the RRBs, commercial banks, and the co-operative banks will go a long way

Table 8.7

Distribution of the Suicide Case Households by Size-class across Credit Amount

(No. of cases)

Category	Up to Rs 25,000		Rs 25,000 to Rs 50,000		Rs 50,000 to Rs 1 lakh		Above Rs 1 lakh	
	Institutional	Non-institutional	Institutional	Non-institutional	Institutional	Non-institutional	Institutional	Non-institutional
Up to 1 acres	3	4	0	1	0	0	0	0
1–2 acres	8	3	3	8	1	4	2	3
2–4 acres	9	1	7	6	3	5	1	5
4–10 acres	7	4	4	4	2	5	4	4
Above 10 acres	0	1	2	1	5	1	0	5
Total	27	13	16	20	11	15	7	17
	(67.50)	(32.50)	(44.44)	(63.56)	(30.55)	(69.45)	(29.16)	(70.84)

Notes: One person borrows from more than one source of credit; figures in parentheses are percentages.
Source: Deshpande (2002a).

in reducing dependence on high cost non-institutional sources and minimizing stress on the farmer borrower. It is essential to simplify the procedure of repayment and create incentive for prompt repayment in order to keep the banking discipline and at the same time, help the farmer by proper counselling.

POLICY DIRECTIONS

The policy formulation and implementation process in Karnataka is different as compared to the other regions with similar constraints. Even though the policy instruments and institutions are similar across states, the policy initiatives are handled differently in Karnataka. Historically, the state polity and bureaucracy has responded quickly in the face of any emerging crisis, as is evident through the initiatives taken by the state so far. Karnataka was the first state to recognize the impending crisis in the agricultural sector and appointed a committee that went into the causes of stagnation and provided far-reaching policy tools (Government of Karnataka 1993). The state also responded to the need to understand the changes due to participation in the World Trade Organization (WTO), and the marketing failures were noted well before the crisis of late 1990s. It is regarded as one of the forerunner states in ushering the process of liberalization. But at the same time, the impending agrarian crisis in the state has also not missed the sight of the policy makers. The state has issued its second Agricultural Policy Document in 2006, dealing with many issues, when some of the other states do not have any such document.

Though the state has a well acknowledged history of responding to the needs of the farming sector, since the mid–1990s there has been a decline in the budgetary allocation to agriculture. Similar is the story of capital formation in the sector through public and private sources. Presently, there are indications of investments into a few regions of the state in the form of hi-tech agriculture and ventures such as food processing and horticultural exports. There is also a need to increase investment to backward regions to create basic infrastructure and technology suitable to their needs.

It is essential to stall the stagnation and improve the rate of growth of the agricultural sector. The net income of the farmer has to increase in order to make agriculture a viable source of livelihood. Further, it is necessary to correct the growing sectoral and regional imbalances during the last two decades. Towards this end, it is necessary to target about 4.5 per cent annual rate of growth in the next few years. In order to achieve this growth rate, it will be necessary to increase the pace of investment in agriculture, both from the private and public sector. The public sector investment should grow at 8 per cent per annum, and the private sector investment almost at the same rate or slightly higher than that. The budgetary support for the agricultural sector has been declining over the years. This puts the sector behind the other sectors, compounded by the fact that the average income in the agricultural sector is quite low compared to the other sectors. Therefore, the budgetary support required would be above Rs 20 per farmer, or in terms of Gross Cropped Area at about Rs 1200–1500 per hectare.

New Policy

There can be no disputing that the agricultural sector of Karnataka needs revitalization. Recognizing the need for a quick policy intervention, the Government of Karnataka came out with the New Agricultural Policy (Government of Karnataka 2006). Among many aspects covered in the policy document, four components assume importance in the present context. First, the agriculturally stagnant areas in Karnataka need to be revived. The districts of northern Karnataka are almost synonymous to backwardness. Unfortunately, adequate attention is not given to rainfed farming in the semi-arid regions of the state. The Agricultural Policy 2006 promised to organize a Rainfed Agricultural Commission, under the chairmanship of a well-known public figure connected with the development of agriculture and allied activities. This commission is expected to draw up plans for the rainfed agriculture of the state in collaboration with the two agricultural universities along with the University of Veterinary Sciences and Fisheries to monitor its implementation through the Department of Agriculture, Government of Karnataka (2006).

Under the policy, a few other initiatives include promotion of crops and varieties in the pockets of modernized agriculture that are suitable for the rainfed regions will be identified, and specific research efforts will be made to direct research towards short duration and drought tolerant varieties. As groundwater shortage is an important constraint in the rainfed areas, it would be necessary to introduce a scheme of water markets and water sharing by 'groups of farmers'. These schemes would be supported by the state, so that unabated groundwater exploitation would be reduced and at the same time, environmental problems be avoided. Extensive efforts are planned for identifying and rejuvenating groundwater recharge zones. Recharge efforts will include recharge points, afforestation in the recharge grails, percolation tanks, and fissures in the aquifer.

The focus of Karnataka's New Agricultural Policy is to achieve at least 4.5 per cent rate of growth in the net income of the farmer through value addition and agro-processing. The new income generation schemes include a rural industrialization programme, as part of a strong linking of agricultural growth with other sectors of the economy.

Agrarian distress is a widespread phenomenon, covering all the farmers almost equally. But the distress is sharply felt by the upwardly mobile farmers across the size-classes as they invest in new ventures. The present crisis not only causes suffering of the farming community but also threatens to wipe out enterprise in the farming sector (Rao 2005). At times, continuous failures faced by this group of farmers inflict severe stress, which can lead to suicidal tendency.

The New Agricultural Policy also envisages addressing the problem of marginalization of the size of holdings, by organizing farmers into *Pragatepar Raitha Okkuta* (PRO), 'Farmer's Consortia'. Under PRO, 10–25 farmers will be encouraged to organize themselves as a group on the lines of self-help groups (SHGs). Gram Panchayat would help to constitute farmers' groups. Each group would purchase inputs as well as sell their produce in the market collectively. Each PRO will be extended 75 per cent subsidy subject to a maximum of Rs 1.10 lakh for the purchase of small machinery and equipment like sprayers, dusters, harvesters, threshers, sellers, and decorticators. The continuing immiserization of poor peasantry needs immediate attention, which can only be achieved by pooling them into small groups to buttress their economic viability and bargaining strength to meet the imperfections in the factor as well as product markets. If these policy initiatives result in institutionalizing a process of sustained high productivity growth in the agricultural sector of the state with special attention to the northern region, then there is a possibility farmers' distress in Karnataka becoming history. Otherwise, it would remain a burning socio-economic and political issue.

Notes

1. The study was conducted by R. S. Dehspande, Agricultural Development and Rural Transformation Centre of the Institute for Social and Economic Change (ISEC), Bangalore, for the Veeresh Committee, which was appointed by the Government of Karnataka to enquire into farmers' suicides in the State (Deshpande 2002a).

2. The committee was appointed in 2000 under the chairmanship of Prof. D. M. Nanjundappa, and the report was submitted in 2002.

References

Aziz, A. and S. Krishna (1996), *Regional Development: Problems and Policy Measures*, New Delhi: Concept.

Dadibhavi, R. V. (1982), 'An Analysis of Inter Taluka disparity and backwardness in Karnataka State, 1975–76', *Indian Journal of Regional Science*, Vol. 24, No. 2, pp. 166–73.

Deshpande, R. S. (2002a), 'Farmers' Distress in Karnataka: A Research Report', Bangalore: ADRT Unit, Institute for Social and Economic Change.

—— (2002b), 'Suicides by Farmers in Karnataka: Agrarian Distress and Possible Alleviatory Steps', *Economic and Political Weekly*, Vol. 37, No. 26, 29 June, pp. 2601–10.

Deshpande, R. S., M. J. Bhende, P. Thippaiah, and M. Vivekananda (2004), *State of the Indian Farmer: A Millennium Study—Crops and Cultivation*, Vol. 9, New Delhi: Academic Foundation.

Deshpande, R. S. and R. V. Dadibhavi (2005), 'Regional Imbalances and Economic Growth In Karnataka', Paper presented at the National Seminar on Accelerated Economic Growth and Regional Balance, Indian Economic Association, Institute for Studies in Industrial Development and Institute for Human Development, New Delhi, 16–18 September.

Deshpande, R. S. and N. Prabhu (2005), 'Farmers' Distress: Proof Beyond Doubt', *Economic and Political Weekly*, Vol. 40, No. 44 and 45, 29 October, pp. 4663–5.

Deshpande, R. S. and K. V. Raju (2001), *Rural Policy for Growth and Poverty Reduction*, Bangalore: ADRT Unit, Institute for Social and Economic Change.

Government of India (various years), *Agricultural Statistics at a Glance, Directorate of Economics and Statistics*, New Delhi: Ministry of Agriculture.

Government of Karnataka (1993), *Stagnation of Agricultural Productivity in*

Karnataka during 1980s, Report of the Expert Committee (Chairman: T. R. Satish Chandran), Bangalore: Government of Karnataka.

—— (2001a), *Agricultural Census of Karnataka,* Bangalore: Directorate of Economics and Statistics.

—— (2001b), *Statistical Abstract of Karnataka, 2000–1,* Bangalore: Department of Planning and Statistics, Directorate of Economics and Statistics.

—— (2002), *Report of the High Power Committee for Redressal of Regional Imbalances (HPCRRI),* (Chairman: D. M. Nanjundappa), Bangalore, Government of Karnataka.

—— (2005a), *Report of the State Level Agricultural Research Review Committee* (Chairman: J. K. Arora), Bangalore: Government of Karnataka.

—— (2005b), *Economic Survey: 2004–5,* Bangalore: Department of Planning and Statistics.

—— (2005c), *Statistical Abstract of Karnataka,* Bangalore: Directorate of Economics and Statistics.

—— (2006a), *Medium Term Fiscal Plan,* Bangalore: Department of Finance.

—— (2006b), *Agricultural Policy of Karnataka—2006,* Bangalore: Department of Agriculture, Horticulture and Animal Husbandry.

—— (2007a), *Economic Survey of Karnataka: 2006–7,* Bangalore: Directorate of Economics and Statistics.

—— (2007b), *Budget 2007–8,* Department of Finance.

—— (various years), *Karnataka At A Glance,* Bangalore: Directorate of Economics and Statistics.

Nadkarni, M.V. (1988), 'Crisis of Increasing Costs in Agriculture: Is there a Way Out?', *Economic and Political Weekly,* 33(39), pp. A114-A119.

Nadkarni, M. V. and R. S. Deshpande (1983), 'Growth and Instability in Crop Yields: A Case Study of Agriculture in Karnataka', South India, Regional Studies, (U.K.), Vol. 17, No. 1, February, p. 30.

Planning Commission (2007), *Karnataka Development Report,* New Delhi: Academic Foundation.

Rao, V. M. (1992), 'Change Processes in Dryland Communities', *Indian Journal of Agricultural Economics,* Vol. 47, No. 1, January–March, pp. 1–23.

—— (2005), *Poverty Reduction in an Elite Driven Democracy,* New Delhi: Danish Books.

Rao, V. M. and H. G. Hanumappa (1999), 'Marginalization Process in Agriculture Indicators, Outlooks and Policy Implications', *Economic and Political Weekly,* Vol. 34, No. 52, 25 December, pp. A133–A138.

Thorat, S. K. and R. S. Deshpande (1999), 'Caste and Labour Market Discrimination', *The Indian Journal of Labour Economics,* Vol. 42, No. 4, pp. 841–54.

University of Agricultural Sciences (2001), 'Viable Size of Operational Holding in Karnataka', Bangalore: Department of Agricultural Marketing

World Bank (2000), 'Indian Policies to Reduce Poverty and Accelerate Sustainable Development', Report No.1947–1N, 31 January.

—— (2004), *Karnataka Public Financial Management and Accountability Study*, Washington D.C.: The World Bank.

Annexure 8.1

Major Characteristics of Agricultural Sector across States

	Year	Andhra Pradesh	Bihar	Gujarat	Haryana	Karnataka	Kerala	Madhya Pradesh
NSA as % of TGA	ATE 2003–4	36.53	60.54	48.85	79.60	51.65	56.36	47.93
Cropping intensity	ATE 2003–4	121.69	138.80	113.75	177.48	116.61	135.71	128.40
% of irrigated area to NSA	ATE 2003–4	38.11	60.56	31.63	84.04	24.90	17.35	33.53
% of rainfed	ATE 2003–4	61.89	39.44	68.37	15.96	75.10	82.65	66.47
% of drought prone area	1981–2	45.60*	25.03	64.39	37.66	79.88	–	19.73@
% of NIA by								
Tanks	ATE 2003–4	12.92	3.23	0.45	0.01	7.75	12.53	2.11
Canals	ATE 2003–4	34.02	27.37	12.57	47.89	32.70	26.64	17.35
Wells	ATE 2003–4	49.09	65.73	86.55	51.68	46.01	31.03	65.69
Irrigation-intensity	ATE 2003–4	129.43	132.07	125.27	178.67	116.66	112.89	103.00
% of waste land	ATE 2003–4	14.48	2.69	16.05	0.96	6.94	4.40	10.48
% of MF	Census 2001	60.90	84.18	29.41	46.07	45.94	95.16	38.56
% of SF	Census 2001	21.83	9.24	30.19	19.24	26.97	3.41	26.51
% of LF	Census 2001	0.57	0.08	1.48	3.27	1.26	0.05	2.26
% of irrigated-MF	Census 1995–6	62.88	74.39	26.93	46.65	46.31	88.00	33.37
% of irrigated-SF	Census 1995–6	19.42	14.21	26.91	19.69	24.10	7.91	23.71
Outstanding credit of commercial banks (in Rs crore)	ATE 2002–3	7525.42	1485.77	3483.21	–	6565.40	2743.95	3738.61

(Contd.)

(Annexure 8.1 contd)

	Year	Maharashtra	Punjab	Rajasthan	Tamil Nadu	Uttar Pradesh	West Bengal	India
NSA as % of TGA	ATE 2003–4	57.01	84.33	43.80	37.04	69.48	61.59	42.08
Cropping intensity	ATE 2003–4	127.22	185.89	123.83	115.79	153.44	176.48	134.38
% of irrigated area to NSA	ATE 2003–4	16.89	95.35	33.39	50.23	73.70	54.52	39.55
% of rainfed	ATE 2003–4	83.11	4.65	66.61	49.77	26.30	45.48	60.45
% of drought prone area	1981–2	40.24*	–	63.09	64.67	14.43	30.21#	32.89
% of NIA by								
Tanks	ATE 2003–4	0.00	0.00	1.15	18.52	0.54	10.50	3.72
Canals	ATE 2003–4	35.25	23.90	24.79	25.69	22.17	23.52	28.23
Wells	ATE 2003–4	64.75	76.03	73.20	55.24	75.63	57.58	63.14
Irrigation-intensity	ATE 2003–4	131.95	188.24	122.62	117.28	145.44	166.01	138.76
% of waste land	ATE 2003–4	10.00	0.16	28.77	23.98	5.85	1.08	13.36
% of MF	Census 2001	47.31	12.34	31.78	74.39	76.88	80.44	63.00
% of SF	Census 2001	28.11	17.35	20.79	15.60	14.25	14.86	18.88
% of LF	Census 2001	0.69	7.22	7.91	0.29	0.15	0.01	1.02
% of irrigated-MF	Census 1995–6	38.06	17.59	27.25	70.27	73.66	73.15	61.72
% of irrigated-SF	Census 1995–6	29.11	16.35	21.71	16.97	15.52	18.96	18.47
Outstanding credit of commercial banks (in Rscrore)	ATE 2002–3	6322.21	4101.70	4039.23	5315.58	7210.58	1803.25	63,891.37

Note: * indicates a year 1985–6, # indicates year 1977–1, @ indicates a year 1983–4 taken from site of Ministry of Water Resources, Government of India. ATE indicates Average Triennium ending; NIA indicates Net Irrigated Area; NSA indicates Net Sown Area; TGA indicates Total Geographical Area; MF indicates Marginal Farmers; SF indicates Small Farmers; LF indicates Large Farmers.

Source: Agricultural Census, Ministry of Agriculture, Government of India (as mentioned in *www.indiastat.com*).

Annexure 8.2
Stagnation in Productivity of Major Crops in Karnataka

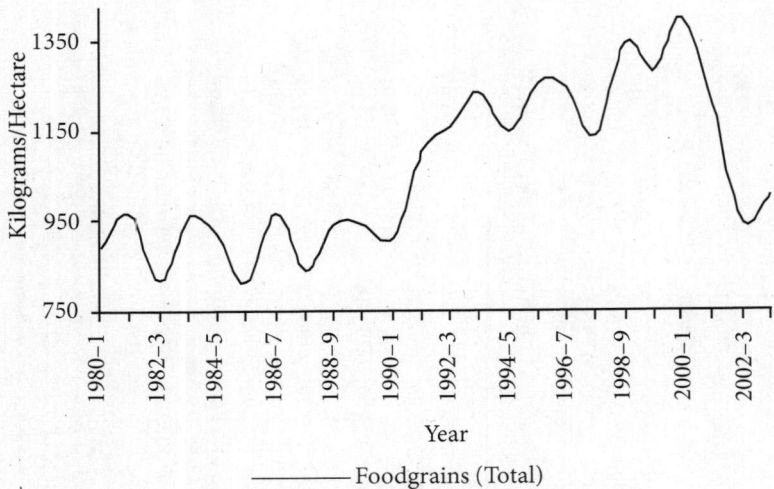

—— Foodgrains (Total)

Yield of Foodgrains (Total) in Karnataka

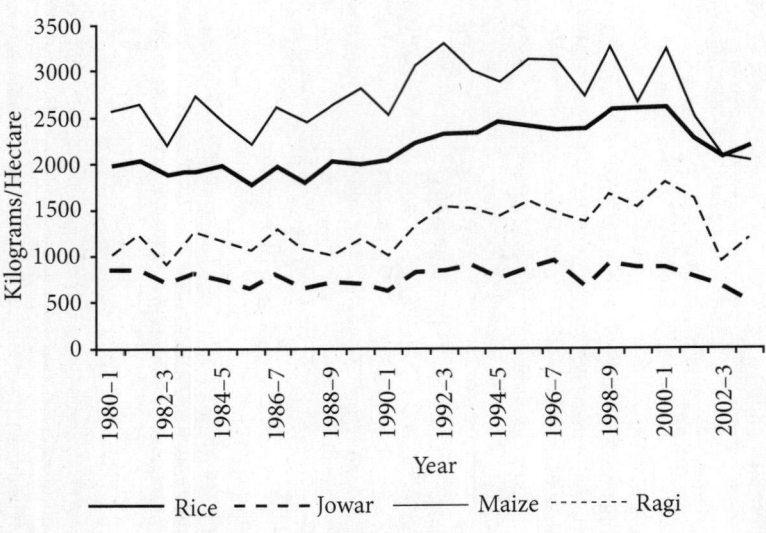

—— Rice - - - - Jowar —— Maize -------- Ragi

Yield of Major Foodgrains in Karnataka

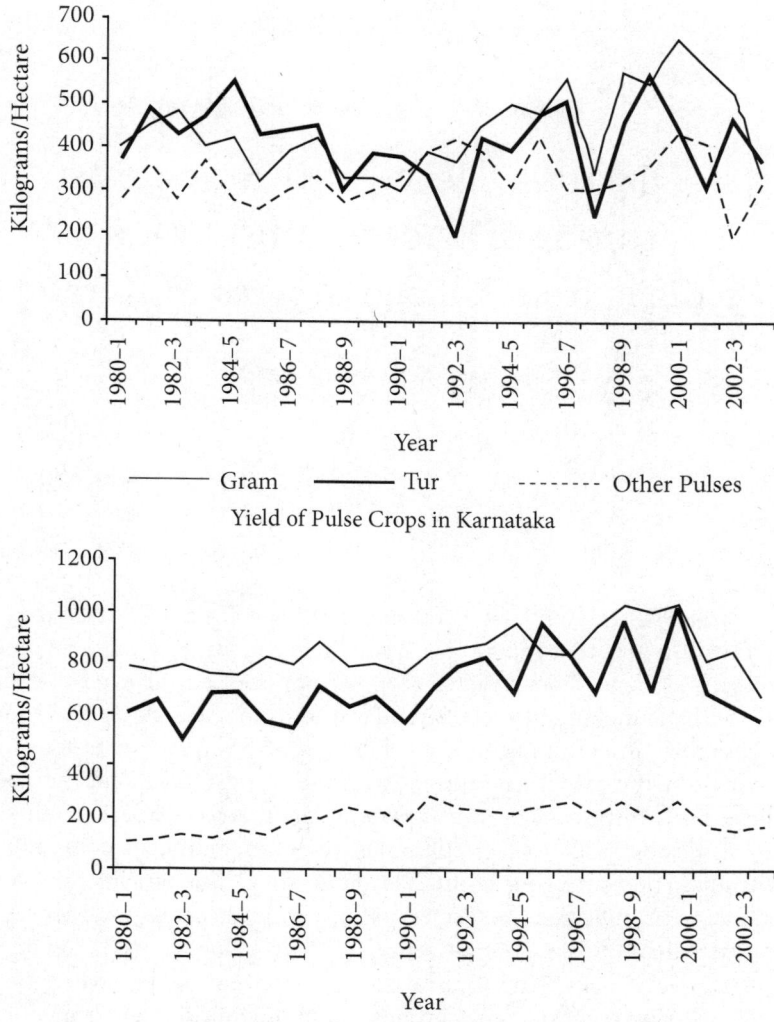

Yield of Pulse Crops in Karnataka

Yield of Cotton, Sugarcane, Groundnut in Karnataka

Note: For sugarcane, yield is calculated in hundred kilograms per hectare; for all others it is kilograms per hectare.

Source: Agricultural Statistics at a Glance, Ministry of Agriculture, Government of India, various years.

9

Distress, Debt, and Suicides Among Farmer Households
Findings from Village Studies in Kerala[1]

K. N. NAIR AND VINEETHA MENON

INTRODUCTION

Agrarian distress and farmers' suicide is a subject of widespread discussion in recent years.[2] It is argued that the primary cause of the distress is the neoliberal policies followed by the Central government since the launching of the economic reforms in the early 1990s. The belief that agriculture would continue to grow even after withdrawal of state support, by integrating it with the world market, appears to have been disproved by the experience of the recent past.[3] Going by indicators such as agricultural investment, output, prices, and incomes, the emerging picture is depressing.[4] The gravity of the situation is reflected in farmers' suicides in those regions where farmers have taken up cultivation of high-value crops involving high level of risks.[5] The Central and state governments have come out with intervention programmes for mitigating the sufferings of the people in the affected areas. These programmes include measures for debt relief, and specific schemes for generating employment and income for the affected. However, while framing of such policies and programmes, the factors and processes underlying the agrarian distress remain unaddressed. Part of the reason for this lacuna is the lack of region-specific empirical studies that examine the causes of

Figure 9.1: Location of Villages Selected for the Study in Kerala

Note: Map not to scale.

Source: Prepared using 'Census Info 1.0, IND_CEN01', Census 2001, Office of the Registrar General and Census Commissioner, Ministry of Home Affairs, India, 2001.

distress, its consequences on the household livelihood strategies, and accumulation of household debt.

This chapter is an attempt to put together some of the insights gained from a study of the nature and dimensions of agrarian distress in three villages situated in Wayanad and Idukki districts of Kerala. Wayanad in north Kerala has faced severe agrarian crisis since 2000.[6] This is reflected in a large number of suicides being reported from among the farming community. According to the State Farmers' Debt Relief Commission,[7] there were 371 suicides by farmers in the district during the period 2000–6.[8] Idukki district in central Kerala, the second district surveyed, has also been under the grip of agrarian distress and reported 89 cases of farmers' suicides during the same period. Both the districts are located in the highland region of the state and agriculture and plantations contribute to roughly 35 per cent of the district income, with about 50 per cent of the labour force absorbed in these sectors. A perusal of the various district level development indicators show that these are two of the backward districts in the state that still retain their agrarian characteristics.[9]

In order to unravel the factors and processes underlying the agrarian distress, on the one hand, and indebtedness and suicides among the agrarian population, on the other, intensive research was conducted in two villages in Wayanad and one village in Idukki.[10] The villages selected for study in Wayanad were Kappikkunnu in Pulpalli Panchayat and Cherumad in Nenmani Panchayat (Figure 9.1). The choice of these three villages was based on the diversity they present within the plantation-based agriculture. Pulpalli Panchayat has reported the maximum number of suicides in Wayanad in recent years, and this also happens to be a location where the increase in coffee and pepper prices in the 1990s had generated a booming economy. On the other hand, Nenmani Panchayat witnessed increasing agricultural prosperity and did not record many suicides following a fall in prices since late 1990s. The village selected for study from Idukki is Upputhara in Upputhara Panchayat. One of the characteristics of this panchayat is that there exist in it both small agricultural holdings and large estates. Unlike other highland regions in the state, farmers in Upputhara have taken up cultivation of tea as one of their crops along with pepper, cardamom, and various food crops. The crash in prices since late 1990s of pepper, cardamom, coffee, and tea has adversely affected the economy of the panchayat.

The survey involved, first, listing of all households in the selected villages to collect information on the principal source of income of the household and the nature of ownership and the extent of operational landholding. Based on the data, a stratified random sample was chosen giving due weightage to both sources of household income and land owned.[11] Apart from the household survey, the study also made use of qualitative research techniques such as focus groups, semi-structured interviews, and case studies. Detailed description of the study locations, methodology, and results from the study for each of the locations is available as separate papers (Nair, Paul and Menon [2007], Nair, Ramakumar and Menon [2007], and Nair, Vinod and Menon [2007]). The first round of household survey and qualitative investigation was started in Cherumad in June 2003, and completed in Upputhara in December 2004; taking roughly about six months in each village. Because of this, the qualitative estimates have limitations for purposes of comparison. However, the final round of research using Participatory Rural Appraisal (PRA) tools was conducted in the villages from July to September 2006, and the information thus collected supplemented the earlier data and made comparison of the changes taking place across these villages possible. Based on these surveys and the individual study reports, this chapter provides a comparative analysis of agrarian distress, indebtedness, and suicides among the agrarian population in the study areas. The second section of the chapter examines the causes of agrarian distress in the study villages and the impact of distress on household livelihoods; the third and fourth sections discuss different dimensions of household indebtedness. The issue of 'suicides' is taken up in the fifth section. The last section brings out the implications of the findings for the design of mitigation measures.

Causes of Agrarian Distress

For understanding the causes of agrarian distress, factors that are exogenous as well as endogenous to the study area are taken into consideration. Factors such as trade liberalization, other macro-economic policy changes, and weather-related conditions fall in the first category whereas local ecology, social structure, and institutional arrangements fall in the latter category. The period since 2000 [the post-World Trade Organization (WTO) period] has witnessed a sharp fall and an increased volatility in the prices of a number of important crops such as pepper, coffee, cardamom, and tea (Subrahmanian 2007). The fall in prices was sharper in the case of pepper and coffee than in tea and cardamom. Domestic prices of these commodities have moved more closely with international prices, confirming the contention that domestic prices are increasingly shaped by international prices. This period has been also marked by sharp fall in rainfall and increased incidence of drought that adversely affected the crop yields. These factors together have resulted in a decline in farm incomes, adversely affected the livelihood of the agrarian population, and created conditions of distress.

A comparative understanding of the important socio-economic characteristics of the study villages will be useful to comprehend the impact of agrarian distress and for an analysis of the indebtedness that follows. Agriculturally, the selected villages are dependent on plantation crops and the average size of holding is in the range of 1.4–1.7 acres. The similarities end there (Table 9.1). In terms of other socio-economic characteristics, there is a wide variation between the Wayanad villages and Idukki village. Occupationally, 81–82 per cent are dependent on agriculture in the Wayanad villages, whereas it is only 63 per cent in Upputhara. The Scheduled Tribe (ST) households account for 27–29 per cent in the Wayanad villages, whereas there are no ST households in Upputhara. Upputhara, because of dependence on relatively higher proportion (37 per cent) of workforce in the non-farm sector, where average incomes are higher and less uneven, appears to have relatively less of general distress, as can be seen from the proportion (2.2 per cent) of households with monthly per capita income (MPCI) of less than Rs 500. Notwithstanding the differences, agricultural distress is common amongst all the three villages.

Since the year 2000, all agricultural households in Wayanad villages and 97 per cent of the agricultural households in Upputhara reported a fall in incomes. Decline in prices and fall in yields were reported as the major reasons. Fall in yields are attributed to two main reasons, viz. adverse weather conditions and pests. While adverse weather conditions figure as an important reason among households both in Wayanad (50 per cent) and Upputhara (35 per cent), pests are cited as

Table 9.1

Characteristics of Households Selected for the Study

Indicator		Cherumad	Kappikkunnu	Upputhara
1	Estimated number of households	445	684	420
2	Average household size (Nos.)	4.5	4.5	3.9
3	Occupational structure:			
	Cultivation	39.8(177)	30.8(211)	32.6(137)
	Agricultural labour	41.8(179)	50.1(343)	30.3(127)
	Labour other than agriculture	5.8(26)	11.4(78)	19.8(83)
	Regular salaried & wage paid employment	8.7(39)	1.6(11)	4.8(20)
	Trade, transport and commerce	3.8(18)	–	2.1(9)
	Forest-based activities	0 (0)	2.8(19)	–
	Others	1.8(8)	3.2(22)	10.5(44)
	Total	100.0(445)	100.0(684)	100.0(420)
4	MPCE (mean)	1,488.93	988.90	1,328.38
5	Per cent of households <Rs 500/- MPCE	25.3	25.7	2.2
6	Average of landholdings (in cents)	166.40	138.70	168.66
7	Per cent of households with <10 cents	17.4	20.2	4.8
	Per cent of SC households	1.1	6	13.1
	Per cent of ST households	27.4	28.6	–
	Per cent of households with electricity	58.3	63.8	65.8
	Per cent of households with latrine facility	71.3	70.5	92.9

Note: Parentheses give number of households.
Source: Field survey, 2004.

an important cause only in the Wayanad villages (40 per cent) but not in Upputhara (8 per cent). The other reason for the decline in agricultural incomes, viz. the decline in prices of agricultural commodities, is attributed to liberalized imports by a majority of households (80 per cent) in the Wayanad villages, while the Upputhara villagers perceived reduction in exports as a reason as much as increase in imports. These perceptions of farmers are corroborated by the available aggregate

Table 9.2
Export of Major Commodities from Kerala

Commodity	Average annual growth (per cent)			
	1990–1 to 1999–2000		2000–1 to 2005–6	
	Quantity	Value	Quantity	Value
Pepper*	17.8	48.5	–11.5	–22.4
Coffee@	13.2	–	–26.8	–25.5
Tea@	26.5	–	–8.0	–18.8

Notes: *Total exports from India, of which around 90 per cent is from Kerala.
@Exports taken place through the Cochin Port.
Source: Jeromi (2007).

Table 9.3
Import of Agricultural Commodities to India

Commodity	Quantity imported average annual growth during 1996–7 to 2005–6
Pepper	33.6
Cardamom	88.8
Coffee	44.5
Tea	30.5

Source: Jeromi (2007).

data of exports and imports of important crops of the state. Table 9.2 shows that exports of pepper, coffee, and tea which experienced high rate of growth during the 1990s, recorded steep decline since 2000–1. In contrast, imports of pepper, cardamom, coffee, and tea into India (Table 9.3) registered an unprecedented growth during the decade beginning with 1996–7 (Jeromi 2007).

In the Wayanad villages, pepper is the main plantation crop, accounting for about 35 per cent of cropped area, followed by coffee (15 per cent) and ginger (6 per cent). In Upputhara, tea (45 per cent) is the main crop, followed by pepper (25 per cent) and cardamom (15 per cent). Although cultivation of tea is done mostly in the estate sector, in Upputhara it has spread to farm holdings also. The farmers' perception of the causes of the fall in agricultural prices is by and large a reflection of their recent experiences with these crops. Pepper, cardamom, coffee, and tea are primarily export-oriented crops, for which domestic prices were affected by import policies. This fact is reflected in the farmers' responses.

Since decline in crop yields is reported as a major reason for the fall in agricultural incomes, it is worthwhile to explore the trends in productivity of major crops grown in the study villages. Since village level estimates of crop yields over time are not available from any source, the yield data was generated through focus group discussions (FGDs) in the study areas with farmer groups.[12] The data (Table 9.4) revealed that though the yield of pepper in general has been higher in Upputhara, it has started declining in all locations, especially since 2000. Coffee also reported a decline in yield, but the order of decline was lower than in pepper.

The yield of rice was about 1600–1700 kg per acre in the 1990s, and it declined to the range of 1400–1500 kg per acre in both the Wayanad villages. The yield of ginger, which centred around 3 tonnes per acre in the 1980s, declined to about 2 tonnes since 2000. In the case of tea, the yield has declined from 1.4–1.8 tonnes per acre in the 1980s to 400–600 kg since 2000. The only crop of this area which did not record any decline during the past two decades is cardamom.

Table 9.4

Changes in the Productivity of Pepper and Coffee in the Study Villages

Period	Range of productivity (kg per acre)		
	Cherumad	Kappikkunnu	Upputhara
	Pepper		
1980–90	500–600	900–1000	1000–1100
1990–9	200–300	750–900	800–900
2000–5	60–80	200–300	300–350
	Coffee		
1980–90	400–500	450–500	475–500
1990–9	300	250–300	250–300
2000–5	200–300	250–300	275–300

Source: Field survey, 2004–5.

The decline in productivity of important crops will have to be viewed along with the changes in their prices. The cause for concern is that along with decline in yields there has also been steep decline in prices of these spices since 2000. The price of pepper, which was around Rs 80 per kg during the first half of the 1990s, increased to a peak level of about Rs 260 per kg during the second half and centred around Rs 180 per kg in 2001 and 2002; it showed a sharp fall in the subsequent years, reaching a level of Rs 60–70 per kg by 2004–5. This crop accounted for more than one-fourth of the area under cultivation in all the three villages and the levels of productivity are relatively high (especially in Upputhara and Kappikkunnu) when compared to the state or the district average. Under conditions of upswing in prices, income generated from this crop would have been substantial. However, the decline in both yields and prices since 2002 has resulted in significant fall in incomes from pepper cultivation. The price of coffee, the second important crop in the Wayanad villages, also showed a similar trend to that of pepper. In the 1990s, deregulation of coffee procurement by the Coffee Board, and the rapid expansion of private trade and its closer integration with the world market, have resulted in significant upward movement in prices. But this process of integration has resulted in downswing in prices since 2000. Like pepper, coffee has also added to the prosperity of farmers in Wayanad villages in the 1990s; the fall in prices and yields since 2000 has aggravated the fall in incomes. In Upputhara, cardamom has emerged as an important crop and, given the fact that its yield did not show any decline though prices have fallen, the price factor would have created lesser impact than in the case of pepper and coffee. However, the fall in prices of tea has severely affected the incomes of the small growers. Cultivation of ginger would have brought in some degree of prosperity in all the three villages in the 1990s, since this crop had good yield and attractive prices. However, in recent years its prices have declined, thereby contributing to a further fall in agricultural incomes.

The persistence of adverse weather conditions for a few years consecutively has resulted not only in a decline in pepper yields, but also in decay and death of pepper vines and some of the commonly planted support trees. The increase in the incidence of pests and diseases has contributed to increase in the quantity of pesticides used for maintaining yield levels. Since there is enough evidence to suggest that the costs of inputs in agriculture have been increasing at a faster rate than prices of agricultural commodities in general, the fact that both the prices and yields of a number of crops that contributed to a good part of the incomes of agrarian households in the study areas have shown sharp fall in the first half of the present decade would have created conditions of distress, shocks, and vulnerabilities among farmers.

In the absence of data on expenditure and income from farm business, it is not possible to bring out clearly the extent to which the decline in agriculture has affected the economics of farming in the study villages.[13] However, another study (Kurup 2006) conducted in a village adjacent to Kappikkunnu during the year 2003–4 provides data on the income/loss per holding from cultivation. These data are given in Table 9.5. It may be noted that even this survey did not collect data for each activity of farm business separately; it is an overall figure reported by the sample holding. The data show that during the reporting year, the sample holdings reported an average loss of Rs 1689. Only holdings with one acre and below reported some profit, possibly because of intensive use of family labour. The reported figures did not include the cost of family labour, and if that were also included, not only would the magnitude of loss increase, but even the lowest size group might also turn out to incur loss.

Table 9.5

Expenditure and Income in Farm Business by Size of Holding in Pulpalli Panchayat

Variables	Size of holding			
	<1 acre	1–3 acre	>3 acre	Total
No. of sample holdings	140	50	60	250
Household size	3.9	4.4	3.0	3.82
Average size holding	2.3	1.6	4.5	1.6
Farm expenditure per holding	1,269.9	7,760	13,000	5,383.2
Income from farming per holding	1,803.3	4,620	733.3	3,693.8
Income/loss per holding	533(+)	3,140(−)	1,689(−)	1,689(−)

Source: Kurup (2006).

The decline in agricultural income has adversely affected not only the cultivators, but also agricultural labourers. The data gathered through FGDs in the Wayanad villages show that the money wages of both male and female labourers have recorded increase during the 1990s, but has shown a fall during the period of price crash, a trend from which this sector recovered only by 2006 (Table 9.6). In Upputhara, however, the money wages did not decline at all. It stood at around Rs 70 to Rs 100 during the period 1995–2000 for male agricultural labourers and around Rs 50 to Rs 80 for female agricultural labourers. The rates increased to Rs 125 for male and Rs 100 for female by 2006. The reasons for the differential trends in wages between Upputhara and the Wayanad villages are that the workers in Upputhara have been able to find jobs elsewhere. In both the Wayanad villages, the number of days of employment for male agricultural labourers, which was about 150 to 160 days a year in 1990, had fallen to 80 to 90 days a year by 2006. In the case of female labour, from about 170 days a year in the 1990s, it declined by one half. The fall in the days of employment along with decline of wage rates resulted in the fall of annual earnings per worker. With the decline in agriculture, non-agricultural employment largely related to trade and transport of agricultural commodities and other service-related activities also declined, and the wages of

Table 9.6

Trends in Money Wages of Agricultural Labourers in Kappikkunnu

Year	Wages (Rs)	
	Male	Female
1970	1.50	0.75
1981	13	7
1990	60	30
1995	70	40
1998	100	80
2000	80	50
2002	90	70
2005	100	80
2006*	100	80

Note: * Wages for head loading of harvested paddy are Rs 150.
Source: Field survey, 2004–6.

workers in these sectors had a pattern similar to that of agricultural wages. Compared to the Wayanad villages, in Upputhara, the income sources are more diversified and, therefore, the impact of shocks on the livelihood of workers has been relatively low.

The fall in incomes of the agrarian population has resulted in such households taking up various coping and adaptive livelihood strategies. Some of the coping strategies noted across villages include: (i) reduction in expenditure on farming, (ii) reduction in household consumption expenditure on relatively costly food items such as meat, milk, fish, and saving on electricity, cooking gas, travel, clothes, and footwear, (iii) shift more towards public services, for example, withdrawal of children from unaided schools to enrol them in government schools, shift from private hospitals to government hospitals for health care, and (iv) migration of adult members to adjacent or distant places in search of work. There has been migration of workers from the Wayanad villages to Tirupur in Tamil Nadu to work in cotton mills; agricultural workers from the region have moved to Coorg in Karnataka to practise ginger cultivation; migration of female members of households in Upputhara to various places in the adjacent districts as home nurses; and women workers from all the villages migrating to other regions as housemaids. Some of this migration now turns out to be adaptive strategy, since they remain in their new places of work for long periods. Other important adaptive strategies developed by households for generation of income and employment include diversification of cropping pattern, for example some cultivating vanilla and some shifting towards food crops such as tubers, vegetables, and banana. The scope for diversification seems to be more in the medium and large holdings than in the small and marginal holdings because of land area constraints. The degree of crop diversification is seen to be higher in Upputhara than in Kappikkunnu and Cherumad. Many households have been taking up dairying as a supplementary source of income. Poor households have been taking up lease farming as a means to augment their income. Apart from individual lease arrangements, leasing of land by self-help groups (SHGs) of women has also been taking place in the study villages. There is growing awareness among farmers on the adverse consequences of the use of chemical fertilizers and pesticides, and a change towards organic cultivation.

Some of the measures taken by the state, either directly or through local self-governments, have been useful in giving some relief to the affected farmers. These measures include: (i) the public distribution system that provides free ration to the ST households during periods of intense distress, and the supply of rice at lower prices to below poverty line (BPL) households are measures that have prevented chronic hunger and starvation among the poor; (ii) though there has been reduction in public expenditure on health care and education, the transfer of some of these functions to the local self-governments has resulted in greater participation of the local communities in the management of these institutions and improvements in their functioning, and (iii) expansion of certain other basic amenities like supply of drinking water, housing and sanitation to the poor by the local self-governments has also provided some relief to the poor. However, none of these measures could contribute significantly to mitigate the fall in incomes and livelihoods of the people. Households have to borrow from various sources to meet their expenses on consumption and investment, and delay the repayment of their pervious loans, as a result of which debts have accumulated among the households.

HOUSEHOLD INDEBTEDNESS

Taking into account the specific situations prevailing in Kerala's villages, the enquiry into the sources of loans was modified to include *chitties* and *kuries*, gold loans, formal sources such as cooperatives and banks, and informal sources such as moneylenders, friends, and relatives. Coming to the purpose of loan, apart from expenditure on production (both capital and current expenditure on farm business), other reasons such as repair of houses, acquisition of land and other productive assets, and repayment of existing loans were also included in the survey. The survey collected data on the year of availing loans, the amounts repaid, and the balances outstanding. However, it did not collect separate information on interest rates and amounts of principal and interest on loans outstanding. Information gathered from the banking institutions in the study areas and interviews conducted with banking officials were used to substantiate/cross-check the reliability of the data collected. Since all the quantitative dimensions of household debt may not have been captured in the sample survey, qualitative methods including case studies of indebted

Table 9.7

Average Amount of Debt by Main Source of Household Income in the Study Villages: 2004

Occupation	Cherumad		Kappikkunnu		Upputhara	
	Mean Debt	Household % (No.)	Mean Debt	Household % (No.)	Mean Debt	Household % (No.)
Cultivators	53,508.46	48.6(171)	142,356.63	39.3(211)	92,212.85	31.9(126)
Agricultural labourers	8,928.95	27.5(97)	17,857.14	39.4(212)	24,912.61	30.8(122)
Non-agricultural labourers	23,400.00	7.4(26)	22,361.43	14.5(78)	15,535.29	21.0(83)
Business	80,707.81	4.3(16)	–	–	28,687.50	2.2(9)
Employed	48,521.15	11.0(39)	62,500.00	2.1(11)	157,218.67	3.0(12)
Others	3,000.00	1.1(4)	21,631.28	2.8(26)	57,741.00	11.1(44)
Total	39,046.18	100.0(352)	68,505.13	100.0(537)	52,135.01	100.0(396)

Source: Field survey, 2004.

households were also done to bring out the dimensions and processes of indebtedness not otherwise captured.

The average amount of outstanding debt per indebted household is the highest in Kappikkunnu (Rs 68,505), followed by Upputhara (Rs 52,135) and Cherumad (Rs 39,406).[14] The average debt per household by the main source of household income showed some degree of variation (Table 9.7). In the Wayanad villages, cultivators and agricultural labourers constitute about 75 per cent of the indebted households. In Upputhara, the corresponding figure is lower, about 60 per cent. All the cultivators in Cherumad and Kappikkunnu reported debt, whereas, in Upputhara, the corresponding percentage was about 90 per cent.[15] Among agricultural labourers, about 50 per cent reported indebtedness in Wayanad villages, whereas 90 per cent reported indebtedness in Upputhara. In sum, a vast majority of households in the category of the self-employed in agriculture, and those dependent on agriculture directly or indirectly, are in debt. The amount of debt by social groups shows that in all the three villages, the intermediate castes such as Ezhavas, Chettys, and Muslims, and the forward communities such as Christians and Nairs reported

Table 9.8

Source-wise Distribution of Indebtedness in the Study Villages: 2004

Source of agency	Distribution of Rs 1000 by source		
	Cherumad	Kappikkunnu	Upputhara
Scheduled bank	400	270	163
Cooperative bank	224	345	268
Government financial institutions	55	136	141
Friends and relatives	151	61	103
Chitties/Kurries	9	12	44
Gold loan	148	101	130
Moneylenders	–	29	85
Self-help groups	13	–	–
Others	–	45	65
Total	1000	1000	1000

Source: Field survey, 2004.

higher amounts of debt, than STs such as Paniyar, Kattunayakar, and Kurumar. In numerical terms, Ezhavas and Christians together constituted a high proportion of indebted households. Since the tribals are the poorest among the social groups, and they have practically no assets to offer as collateral, one would expect lower incidence of indebtedness among them.

Data on the amount of debt (Table 9.8) outstanding by sources shows that access to formal sources of credit is higher in the Wayanad villages than in Upputhara. Moneylenders and 'gold loans' are still prevalent in all the villages. The relative importance of different sources of credit in the outstanding amounts of debt is evident from the pattern of distribution per Rs 1000 of outstanding debt, by sources. The estimate reinforces the dominant role of formal credit;[16] they account for 68 per cent of the debt in Cherumad, 71 per cent in Kappikkunnu, and 57 per cent in Upputhara.

The purpose for which debts were incurred is evident from Table 9.9. In Cherumad, 42 per cent of the debt was incurred to meet the

Table 9.9
Purpose-wise Distribution of Outstanding Debt in the
Study Villages: 2004

Purpose	Distribution of Rs 1000 of debt		
	Cherumad	Kappikkunnu	Upputhara
To meet increased production expenditure	424	656	353
To meet increased non-production expenditure	101	55	92
To repair house	235	150	107
To acquire production assets	56	13	-
To acquire land	21	1	43
Medical expense	47	12	56
Marriage expenditure	50	26	100
Repay existing loans	43	3	187
Others	23	86	62
Total	1000	1000	1000

Note: Sample weight applied.
Source: Field survey, 2004.

current and capital expenditure in farm business. In the other two villages, the share of this item is much higher. Repair of houses constituted another important item of debt. The remaining amount was incurred for a variety of other purposes.[17]

CHANGES IN INDEBTEDNESS

Development of credit institutions in the study villages is closely linked with the evolution of cultivation of plantation crops. Before in-migration of population into the study areas from the plains began, landed households had only the facility to exchange rice with traders in the local markets (*chantas*) in exchange for condiments and household utensils. In the colonization scheme in Wayanad, there was a component of agricultural loan that was not repayable but was to be used exclusively for the cultivation of plantation crops. A formal institutional facility for disbursing loans was initiated mainly due to the efforts of in-migrant farmers. They started *Aikya Nanaya Sangham* as a self-help attempt. In Cherumad, a Sangham was floated in 1952 for the whole region by a number of enterprising farmers collecting contributory shares from among themselves and loans were distributed according to the requirements. In 1962, this association was registered as a cooperative bank under the District Cooperative Bank, then functioning with financial assistance from the state government. The banking activities were scaled up, as a result, disbursal of agricultural and non-agricultural credit began as a regular banking process. A board of directors elected by the shareholders, on political lines, handles the day-to-day affairs of the bank. This makes the bank accountable for and sensitive to the requirements of the local people and local interests. Besides, the Coffee Board provided some credit facilities to the coffee growers till 1994, when the Board was dismantled. The households in Kappikkunnu reported that there are three cooperative banks, three nationalized banks, and two private banks functioning in that area, from all of which they avail credit. In Cherumad also, all these types of banks are in operation. Private financial institutions, such as *chit* funds and moneylenders, have become active in the area during the past two decades. In Upputhara, a cooperative credit society started functioning from as early as the 1940s. Apart from the Upputhara service cooperative society, which was formed in 1949 and covers the entire Panchayat, two more cooperative credit societies have started functioning in 2005. Besides, a branch of the Idukki District Co-operative Bank and the Malanad Service Co-operative bank have also been providing services in Upputhara.

Since the 1990s, all the three areas witnessed rapid increase in availing of credit. The institutional sources were the major sources of borrowing up to 2000, but in the later period dependence on non-institutional sources increased substantially. For example, as Table 9.10 shows, the percentage of households that took loans from both formal and informal sources in Cherumad, has grown rapidly since 1995. But, informal sources, which never exceeded more than 30 per cent until the year 2000, reached two-thirds during the period after 2000. Private bankers who issue gold loans and moneylenders are found to have assumed great significance in recent years in Cherumad.[18] Similar tendencies were observed in the other two villages as well.

The rapid increase in the farmers' demand for credit from all sources is linked to an unprecedented price hike in the case of major crops during the 1990s that flared up the aspirations of households. Between 1991–2 and 1996–7, the average annual growth in farm price (expressed

Table 9.10

Distribution of Households with Loans Outstanding by Period across Sources of Loan in Cherumad

Source of Loan	No. of households by period of loan		
	Before 1995	1995–2000	After 2000
Scheduled banks	41 (75.5)	69 (32.9)	99 (20.2)
Cooperative banks	14 (18.7)	65 (31.0)	44 (09.1)
Government financial institutions	–	13 (06.1)	18 (03.7)
Friends and relatives	–	42 (20.0)	81 (16.7)
Chitties	–	–	9 (01.9)
Gold loan	15 (20.0)	21 (10.0)	154 (31.8)
Moneylenders	5 (07.0)	–	80 (16.5)
All sources	75 (100)	210 (100)	485 (100)

Note: Figures in parentheses indicate the percentage of households to the total. The total includes households borrowing multiple sources.
Source: Field survey, 2004.

in percentages) increased by 22.8 per cent for pepper, 15.7 per cent for tea, and by 9.9 per cent for cardamom (Jeromi 2007). Spurred by the rising prices, farmers planned substantial increase in their expenditure not only on agriculture but also on land, housing, and financial assets. This was also a period when banks were competing with one another to entice farmers into taking loans under various schemes, goaded by a bloated impression of credit-worthiness of farmers, an impression that often proved unfounded as it rested on anticipations of rising crops and yields. The farmers too were persuaded to enter into market speculations often based on expectations rather than on hard facts and reliable information of market trends. And they made use of all the loan facilities available from various sources. They used the loans not only for agricultural purposes, but also for various other purposes— productive and non-productive. House construction was an activity that received the greatest priority in the investment decisions in the case of both poor and the not-so-poor farmers. Thatched roofs were changed to tiled or concrete roofs. Floors were cement-plastered or laid with marble, granite, or mosaic. Investments were made in land (either investment in land improvements or purchase of land), for raising new plantations, and in financial assets. It is difficult to get a category-wise quantitative estimate of such investments. However, rough estimates based on the fieldwork in one village, Kappikkunnu, shows that during 1999–2004 substantial investments were made in housing and financial assets, exceeding investments in land and plantations. It is the households belonging to the top expenditure groups that invested more in all such activities. Besides these investments, there appear to have been also acquisitions of consumer durables, by making maximum use of the flexible lending approach of the credit institutions. Two per cent of the households in Cherumad, 7 per cent in Kappikkunnu, and 5 per cent in Upputhara acquired cars or jeeps by availing loans from banks. Nearly 4 per cent of the houses in Cherumad, 12 per cent in Kappikkunnu, and 6 per cent in Upputhara purchased scooters or motorcycle with bank loans. Similarly, purchase of items such as refrigerator, TV, etc. became widespread. The increased prosperity of the 1990s also resulted in changes in the lifestyles of households. To give an indicator, nearly 8 per cent of the households in Cherumad took telephone connections; the corresponding percentage in Kappikkunnu and Upputhara being

16 per cent and 12 per cent respectively. The use of credit institutions for meeting expenditure on health care and education of children (especially sending them for technical higher education courses such as nursing, medicine, and engineering, to private institutions in places outside the state) also became common in the study areas.

The dreams of sustained prosperity of farmers were shattered when the prices of major agricultural commodities collapsed. Between 1997–8 and 2005–6 the annual rate of growth of prices of pepper decelerated to 2.3 per cent and cardamom and coffee experienced a negative growth of –1.4 per cent and –1.05 respectively (Jeromi 2007). Added to this, the adverse impact of prolonged drought, pests, and diseases on crop yields increased farmers' debt burden. They failed to repay loans. Consequently, overdues mounted up. For instance, loan overdues as proportion of loans disbursed were 7.25 per cent in the Pulpalli branch of the State Bank of India in 2000, and by 2005, the corresponding proportion increased to 54 per cent.[19] A similar trend was noted in other banks as well. The overdue position of banks was more severe in Wayanad than in Idukki.

The lending policies of banks have undergone change in response to the mounting overdues. The banks have reduced their long-term and medium-term loans to agriculture.[20] Priority has shifted from long-term to short-term loans, especially loans with gold as collateral. Banks have also stopped issuing fresh loans to defaulting households and started issuing notices of revenue recovery. Under pressure from farmers' organizations,[21] banks have made attempts to adjust old loans against new loans, which was reflected in the bank records as increase in the supply of credit, but in reality it did not ease the credit constraints of farmers.

Pledging gold had been one of the last resorts in raising loans under conditions of distress.[22] The liberal policy of banks to issue gold loans has facilitated the growth of loans, with gold as collateral. The data available from commercial and cooperative banks in the study regions have shown manifold increase in the amount of gold loans disbursed since 2000. The practice of dowry has been getting widespread in the study villages (as it is in the rest of Kerala) in recent years, and gold is an integral part of the dowry. It is also seen that households had resorted to purchasing gold during times of price boom. In Wayanad, earlier there was no dowry system at all among tribal households. However, they have also started adopting this practice, as 'demonstration effect' of the practice followed by the settler communities, especially the Christians. Gold has become a sure form of savings since gold price has been steadily on the increase. Since the late 1990s, a number of private financial institutions that provide gold loans have come up in the study areas. Some of these agencies also specialize in providing loans to debtors to recover their gold pledged with banks and helping them sell it off. In Kerala, pledging gold is considered infra dig to households, especially gold which has been part of a woman's dowry. During the agrarian crisis in Wayanad, many families are reported to have put off marriage negotiations of daughters and sisters, as lands and crops were losing value, while wedding expenses and price of gold were skyrocketing.[23]

INDEBTEDNESS AND SUICIDES

A recent survey (Shreyas 2007) of more than 300 households in which suicide took place during the period 2002–6 in Wayanad district shows some of the socio-economic characteristics of the farmers committing suicides.[24] Farmers committing suicide were mostly males (90 per cent), from across all caste and religious groups of the district, and the majority (60) of them were in the age group of 41–60 years. They belonged mostly to marginal farmers with one acre or less (59 per cent) and small farmers with one to two acres (27 per cent). Much of their borrowing was for productive purpose. About 59 per cent have taken loans for investment in agricultural activities and 13 per cent for repaying old loans. Although many households (40 per cent) could not specify the immediate reason, 40 per cent did specify growing indebtedness in the face of falling productivity and prices. The majority of households (58 per cent) did not receive any compensation from the government, 32 per cent received an amount of Rs 50,000 each, and the rest Rs 10,000 each. Though there are multiple reasons, indebtedness stands out as one of the main reasons. Another survey conducted in Wayanad district also indicates indebtedness as one of the main reasons for suicide (Kurup 2006).

To obtain a better picture of the manner and the process through which indebtedness, combined with other family and social factors, resulted in suicides among members of agrarian households, several case studies of indebted households were conducted.[25] In these case studies, the complex nature of factors and processes contributing to indebtedness of households was looked into; attempts were made to find an explanation of as to why some households succumb to the stress of indebtedness while others struggle on and survive. For this purpose, we examined the two categories of indebtedness analytically. In the case of farmer suicides, it is very clear that these households had been in such dire straits because of the cumulative impact of various distressing life circumstances. In all cases of suicide from our study areas, it is found that these were people who had no hope of coming out of their crisis situations. On top of the heavy burden of the debt and the difficulties associated with making both ends meet, some calamity or the other such as a terminal or chronic illness of a member of the household or the failure of an expected source of income as in the case of death of dairying cattle, or the death or financial ruin of a sole income earner in the family, had befallen them, driving them to the desperate measure of taking their lives. However, the story of survivors shows that even in such cases of unexpected calamities, the presence of support networks and the degree of success with which some are able to negotiate various institutions are crucial factors that enable even persons so deeply troubled to have the tenacity to fight life's battles as is evident from the discussion of three case studies already indicated in Box 5.3, chapter 5, of this volume.

CONCLUDING REMARKS

Micro-level study of three villages in largely plantation-based districts of Wayanad and Idukki reveal that a sharp fall in prices of crops such as pepper, coffee, cardamom, and tea along with fall in their yield levels due to adverse weather conditions created a situation of 'shock and vulnerability' and agrarian distress in the study areas. Every section of the rural population—cultivators, agricultural labourers, non-farm workers—was severely affected by the distress. The adverse weather conditions were only an additional contributing factor to the policy changes of trade liberalization and declining state support to agriculture that generated the agrarian crisis.

Faced with shocks and vulnerabilities, households developed several coping and adaptive strategies that included reduction in household expenditures, out-migration, and diversification of household activities. Mitigation measures of the state, especially the public distribution system, and interventions by Panchayats in health care, education, and supply of drinking water provided some relief to the affected population.

In the face of a prosperous agricultural economy during the 1990s, households raised loans from different sources either for investment or for acquiring consumer durables. However, the shocks experienced from the year 2000 onwards resulted in non-repayment of loans, and additional borrowing to meet production and household expenditure, thereby increasing household indebtedness. Though formal sources of credit accounted for the bulk of the outstanding debt, there is a growing share of moneylenders and other informal sources. The practice of raising gold loan is seen to be widespread. Though the main purpose of borrowing has been to meet expenditures on agricultural production, the share of other household expenditures, especially on housing, education, and health, are on the rise. There are inter-village differences in the magnitude of distress, coping with indebtedness, and adaptive strategies of households. The intensity of the distress and household indebtedness is higher in Kappikkunnu; the region where this village is located also reports very high incidence of suicides in the district. In the other two villages in which household incomes are more diversified, and social networks much stronger, the distress conditions did not result in suicides. Though there are a number of policy, institutional, and household circumstantial factors, indebtedness is an immediate factor that contributes to the incidence of suicides among members of the farming households. There have also been other unexpected calamities, lack of support networks, and the inability of the victims to cope with such circumstances that drove them to put an end to their lives.

It is evident from these findings that there are multiple dimensions to the agrarian distress. Therefore, debt relief alone would not bring solace to the indebted households. Debt relief is at best only a short-term palliative. What is needed is a long-term strategy to mitigate the distress within an integrated framework in which are embedded policies to promote price stability, ecological sustainability of agriculture, strengthening of formal rural credit, strengthening of support networks, and income and employment generation programmes in the rural non-farm sector. The emphasis should be on strengthening those activities that households have developed as part of their coping and adaptive strategies. Such an integrated approach to the agricultural sector would help in addressing the root causes of the malady, in proactively rejuvenating agriculture, farmer households, and ecology, as well as livelihoods in distress-prone areas. Though relief to deceased families is important, it would only aggravate the agrarian distress in the absence of appropriate policies to improve the agriculture sector.

NOTES

1. This chapter forms part of an ongoing project on 'Economic Globalization and State Decentralization: Coping Strategies of Farm Households in South India', sponsored by National Centre of Competence in Research (NCCR) (North–South), Swiss National Science Foundation, Berne and the Department of Geography, University of Zurich, Switzerland. The authors wish to thank Antonyto Paul, R. Mahesh, C. P. Vinod, R.. Ramakumar, V. N. Jayachandran, M. J. Joseph, and C. S. Krishnakumar who were associated in conducting the individual studies. An earlier version of this chapter was presented in a seminar at the Indira Gandhi Institute of Development Research (IGIDR), Mumbai. Thanks are due to Dr P. D. Jeromi, Assistant Economic Adviser in the Reserve Bank of India for his detailed comments. Prof. P. R. Gopinathan Nair, D. Narayana, N. Krishnaji, and Urs Geiser also commented on this chapter. However, none of them are responsible for the views expressed in this. An earlier version has also been put up as CDS working paper (Nair and Menon 2007).

2. A large number of studies are currently available on farmers' suicides for different parts of the country. Some of the important studies are CES (1998), Mishra (2006), Mohanty and Shroff (2004), Mohanty (2005), Parthasarathy and Shameem (1998), Revathi (1998), Rao and Gopalappa (2004), Gill and Singh (2006), and Vaidyanathan (2006).

3. There are two schools of thought on the impact of trade liberalization and globalization on Indian agriculture. According to one school of thought, it would result in increase in the efficiency in resource allocation in agriculture and increased productivity gains; on the other hand, the opposing school argues that it would result in erosion of food security, inequality in the distribution of incomes, and overall negative impact on the poor. For details, see Bhalla (2004).

4. For a summary of this argument, see Reddy (2006a and 2006b).

5. The states that are severely affected by agrarian distress are Andhra Pradesh, Karnataka, Maharashtra, and Punjab,. Within these states, specific

regions growing cash crops for exports are more badly affected than other regions.

6. For a preliminary analysis of the crisis, see Mohanakumar and Sharma (2005).

7. The Government of Kerala, in 2006, through a bill introduced in the state assembly, set up this Commission with the objective of examining the intensity of agrarian distress in any part of the state and, if necessary, to reduce distress and initiate mitigation measures. The Commission is empowered to prevent the recovery of loans by various financial intermediaries from the affected households and to prevent exploitation of the households under distress by the lenders. The Commission can also direct the government to take up mitigation measures. It has a retired judge of the High Court as the Chairman.

8. Data on suicide mortality, the characteristics of the victims such as age, sex, method of committing suicide, etc., are fairly reliable and are available in published form. However, there are limitations regarding the causes of death and the occupations of the victims. While the victims' households have been provided with financial support and other relief measures by the state, if the cause is 'farm crisis', there is a tendency to classify a suicide in the distress-prone areas as 'farmers' suicide'. The data on the number of suicides committed by farmers are compiled by the state agricultural department and supplied to other agencies.

9. For a comparison of the development indicators at the district level, see the Human Development Report for Kerala (Government of Kerala 2006). In both Wayanad and Idukki, about 48 per cent of the workers are engaged in agriculture, either as cultivators or agricultural labourers. On the other hand, for the state as a whole the proportion is only 23 per cent. If we include also the plantation workers, then around two-thirds of the workers in these two districts would be deriving their livelihood from agriculture (including plantations).

10. In Kerala, unlike in other states, a revenue village is a large administrative unit with an average size of about 5000 households and a population of about 20,000 persons. Therefore, for the purpose of conducting micro-level studies, a ward of a panchayat (the lowest administrative unit) is usually taken up. A panchayat ward may have about 500–1000 households depending on the population density. For the purpose of the present study, villages refer to wards of the panchayats concerned.

11. Multi-stage stratified sampling method has been used to select the sample households. At the village level, all households were stratified on the basis of the main source of income (Table 1). Among the major strata,

Table 1

Distribution of Households by Major Strata (Main Source of Income)
in the Study Villages

Strata	Number of households		
	Cherumad	Kappikkunnu	Upputhara
Cultivators	176	211	136
Agricultural labourers	179	343	127
Non-agricultural labourers	26	67	84
Business, trade	17	13	6
Employed	39	14	20
Others	8	36	47
Total	445	684	420

Table 2

Number of Households Canvassed in the Study Villages: 2004

Strata	Households canvassed		
	Cherumad	Kappikkunnu	Upputhara
Cultivators	36	22	27
Agricultural labourers	35	34	25
Others	28	26	32
Total	99	82	84

cultivators and agricultural labourers were again stratified on the basis
of the extent of landholdings. The number of households selected for the
present study are given in Table 2. Appropriate weights were computed for
the analysis.

12. It may be noted that yield levels given are quantities reported by the
respondents from their memory and, therefore, should be taken as broad
indicators of the levels and directions of change.

13. The household survey conducted in the study villages did not collect
data on farm expenditure and income, which would have given some idea
of the intensity of the distress. At the pilot phase of the survey, we found it
difficult to collect reliable data on cost of cultivation and income for each of
the crops and other activities such as livestock keeping.

14. The mean amount of household debt estimated for the study villages is
higher than the amount estimated by the 59th round of the National Sample
Survey (NSS) for the year 2003 for the state. The estimated debt per farmer

household for Kerala by the NSS survey was Rs 33,907 and for the country as a whole it was Rs 12,585. There exist a few other sources of estimates also for the state as a whole and for specific regions. But these sources of data are not comparable due to different reference periods and differences in the concepts and methodologies used.

15. The incidence of indebtedness among the farmers in the sample villages is much higher than the state and national averages.

16. The findings from the household survey are in conformity with that available from the 59th Round of the NSS. According to the NSS survey, the percentages of households taking credit from co-operatives, banks, and other government sources, were 46, 42, and 8, respectively, for the country as a whole. The corresponding percentages for Kerala as a whole were, 26, 27, and 3.

17. The purposes for which loans were taken in the survey villages differed from those of the NSS estimates for the state as a whole in one important respect. The amount of debt incurred for production expenditure in non-farm business was considerably higher in the survey villages than the corresponding amount reported by the NSS survey for the state as a whole. The NSS data show that about 25 per cent of the outstanding loans were incurred by farmer households for meeting current and capital expenditure in farm business whereas for the country as a whole it was 60 per cent. In Kerala, non-farm business and other expenditures together constitute nearly 50 per cent of the loans outstanding.

18. The activity of moneylenders calls for some explanation. In all the villages, moneylenders from Tamil Nadu had been present even before the price boom began. But, with the price boom, several new local moneylending agencies started flourishing in the area: some of them were large farmers who had accumulated huge surpluses during the period of the boom. They have come to be known locally as 'blades', implying the double-edged annihilating nature of their transactions carrying usurious rates of interest and practising extortionary methods in cases of default. Many farmers raise emergency loans from them since the procedure for getting loans is much simpler with them than with formal credit institutions.

In the recent past, in the study regions in Wayanad, a number of incidents of destroying of the establishments of moneylenders and their loan documents have been reported—measures which the borrowers resorted to as part of their responses to attempts by moneylenders to confiscate properties of defaulting borrowers. Activities of local private moneylenders have come down drastically as moneylenders from Tamil Nadu withdraw from business with fall in prices and for fear of public wrath. In the Idukki region, since the economic situation is better than in Wayanad, the number of formal credit institutions is relatively small and the activities of moneylenders are more widespread. The fall in incomes and the squeeze in the local credit system have also resulted in households increasingly resorting to borrowing from sources such as friends and relatives.

19. The following data supplied by the State Bank of India (Pulpalli Branch) clearly show the types of loans issued and the mounting overdue position of banks.

Table 3
Amount of Loan by Type: 2000–6

(Rs lakh)

Amount of loan by type	2000	2001	2002	2003	2004	2005	2006
Agriculture loans	18.16	27.84	31.22	30.58	50.78	68.7	86.28
SSI loans	2.94	2.94	0.98	3.87	2.01	4.01	6.95
Others	208.93	294.3	359.69	395.87	400.17	376.97	357.28
Total	230.03	325.08	391.89	430.32	452.96	449.68	446.92
Loans overdue	16.68	45.65	96.98	213.29	249.72	242.00	255.38

20. The data collected from the Services Co-operative Bank of Pulpalli show that during the early years of agrarian distress (2000–3), short-, medium-, and the long-term loans issued to agriculture were virtually stagnant; they showed increases since then. A similar trend was also noted in the case of other banks. This trend is not consistent with the state-level data on loan advances to agriculture by commercial banks during the financial years from 1999–2000 to 2005–6, available from the State Level Bankers Committee. Part of the reasons for the increase could be the carry over of overdue loans also, along with actual advances made to agriculture.

21. The Farmers' Relief Forum, which started functioning in 1994, mobilized a revolving fund of Rs 10,000 to aid farmers to renew loans by paying arrears of interest. However, even this forum did not help farmers when they failed in repaying loans and banks initiated attachment procedures. At this juncture, under Farmers' Relief Forum, 'japthi thadayal'—prevention of attachment procedures—was initiated, which turned even violent at times. Farmers say that it is not the act of attachment itself, but the manner in which it was implemented, that hurt the pride of the farmer. The attachment procedure took place with the accompaniment of drum-beating, thus bringing the action to public attention.

22. Gold loan is easy to obtain, since it is accepted readily as collateral by both formal and informal agencies for supply of credit. The rate of interest charged on gold loan by various credit institutions varied in the range of 9.5 per cent to 12.5 per cent per annum. The interest rates charged by moneylenders varied from 24 to 60 per cent, depending on the duration and the type of security given as collateral. The interest rates charged by commercial banks on agricultural loans ranged from 8.5 to 9 per cent. These are the rates that prevailed during the year 2005–6.

23. A panchayat member in the Wayanad district narrated his own experience. In his ward, a family was not able to find the means to go ahead with a marriage that had been already arranged as they could not purchase the promised quantity of gold. He intervened in the situation and persuaded a jewellery owner, who was his friend, to lend jewellery for the occasion so that the prestige of the families concerned could be kept up. The marriage did, thereupon, take place and the jewellery returned to the shop thereafter. This was possible only because the bridegroom's family could empathize with the bride's family and the distress was viewed not merely as the household's problem, but also as a widespread one, shared by the region as a whole.

24. The survey results reported here are taken from a study conducted by Shreyas, a non-governmental organization (NGO) which was working in Wayanad district in 2006. The respondents in the survey were members of the victims' households. The nature of information gathered included personal details of the victim, his/her socio-economic background, and the circumstances in which the victim committed suicide. A more or less similar survey was conducted by Mohankumar and Sharma (2006) in 30 households which had a suicide victim each. Both the surveys showed that the average amount of debt among the households were much higher than the figures reported per indebted households in our survey. The Shreyas survey reported a debt of Rs 72,000, whereas the Mohankumar and Sharma survey reported Rs 793,858.

25. Thirteen case studies from the three villages of the study areas were conducted. Detailed accounts of these case studies are available in Nair, Paul and Menon (2007), Nair, Ramakumar and Menon (2007), and Nair, Vinod and Menon (2007).

REFERENCES

Bhalla, G. S. (2004), *State of the Indian Farmer: A Millennium Study— Globalisation and Indian Agriculture*, Vol. 19, New Delhi: Academic Foundation.

Centre for Environmental Studies (CES) (1998), *Gathering Agrarian*

Crisis: Farmers' Suicides in Warangal District (AP): A Citizen's Report, Hanamkonda (AP): CES.

Deshpande, R. S. (2002), 'Suicides by Farmers in Karnataka: Agrarian Distress and Possible Alleviatory Steps', *Economic and Political Weekly*, Vol. 37, No. 26, 29 June, pp. 2601–10.

George, P. S. (2005), *Globalization and Agricultural Transition in Kerala*, Report submitted to the Indian Council of Social Science Research, mimeo, Thiruvananthapuram: Centre for Development Studies.

Gill, A. and L. Singh (2006), 'Farmers Suicides and Response of Public Policy: Evidence, Diagnosis and Alternatives from Punjab', *Economic and Political Weekly*, Vol. 41, No. 26, pp. 2762–8.

Government of Kerala (2006), *Human Development Report 2005*, Thiruvananthapuram: State Planning Board, Kerala.

Gulati, A., R. Mehta, and S. Narayanan (1999), 'From Marrakech to Seattle: Indian Agriculture in a Globalising World', *Economic and Political Weekly*, Vol. 34, No. 41, 9–15 October, pp. 2931–42.

Jeromi, P. D. (2007), 'Farmers' Indebtedness and Suicides: Impact of Agricultural Trade Liberalisation in Kerala', *Economic and Political Weekly*, Vol. 42, No. 31, 4 August, pp. 3241–7.

Kurup, K. K. N. (2006), 'Agriculture Debt and Suicides in Wayanad' (in Malayalam), *Deshabhimani*, Onam Special Issue, Thiruvananthapuram, pp. 175–7.

Mishra, S. (2006), 'Farmers' Suicides in Maharashtra', *Economic and Political Weekly*, Vol. 41, No. 16, 22 April, pp. 1538–45.

Mohanakumar, S. and R. K. Sharma (2006), 'Analysis of Farmer Suicides in Kerala', *Economic and Political Weekly*, Vol. 41, No. 16, 22 April, pp. 1553–8.

Mohanty, B. B. (2005), 'We Are Like the Living Dead': Farmer Suicides in Maharashtra, Western India', *The Journal of Peasant Studies*, Vol. 32, No. 2, April, pp. 243–76.

Mohanty, B. B. and S. Shroff (2004), 'Farmer's Suicides in Maharashtra', *Economic and Political Weekly*, Vol. 39, No. 52, 25 December, pp. 5599–606.

Nair, K. N. and V. Menon (2004), 'Reforming Agriculture in a Globalizing World: The Road Ahead for Kerala', IP6 Working Paper No. 3. Berne: NCCR-North-South (Swiss National Science Foundation).

——— (2007), 'Distress, Debt and Suicides among Agrarian Households: Finding from Three Village Studies in Kerala', Working Paper No. 397, Thiruvananthapuram: Centre for Development Studies.

Nair, K. N., A. Paul, and V. Menon (2007), 'Livelihood Risks and Coping Strategies: A Case Study in the Agrarian Village of Cherumad', mimeo, Thiruvananthapuram: Centre for Development Studies.

Nair K. N., R. Ramakumar, and V. Menon (2007), 'Agrarian Distress and Rural Livelihoods: A Study in Upputhara Panchayat, Idukki District, Kerala State', mimeo, Thiruvananthapuram: Centre for Development Studies.

Nair K. N., C. P. Vinod, and V. Menon (2007), 'Agrarian Distress and Livelihood Strategies: A Study in Pulpally Panchayat, Wayanad District, Kerala', mimeo, Thiruvananthapuram: Centre for Development Studies.

Nayyar, D. and A. Sen (1994), 'International Trade and Agricultural Sector in India', in G. S Bhalla (ed.), *Economic Liberalisation and Indian Agriculture*, New Delhi: Institute for Studies in Industrial Development.

Parthasarathy, G. and Shameem (1998), 'Suicides of Cotton Farmers in AP: An Exploratory Study', *Economic and Political Weekly*, Vol. 33, No. 13, 28 March, pp. 720–6.

Rao, V. M. and D. V. Gopalappa (2004), 'Agricultural Growth and Farmer Distress: Tentative Perspectives from Karnataka', *Economic and Political Weekly*, Vol. 39, No. 52, 25 December, pp. 5591–8.

Reddy, D. N. (2006a), 'Changes in Agrarian Structure and Agricultural Technology: Is Peasant Farming Sustainable under Institutional Retrogression', in R. Radhakrishna, S. K. Rao, S. Mahendra Dev, and K. Subba Rao (eds), *India in a Globalising World: Some Aspects of Macroeconomy, Agriculture and Poverty, Essays in Honour of C.H. Hanumantha Rao*, New Delhi: Academic Press.

—— (2006b), 'Economic Reforms, Agrarian Crisis and Rural Distress', 4th Annual Professor B. Janardhan Rao Memorial Lecture, Professor B. Janardhan Rao Memorial Foundation, Warangal, Telengana.

Revathi, E. (1998), 'Farmers' Suicides: Missing Issues', *Economic and Political Weekly*, Vol. 33, No. 17, 16 May, p. 1207.

Sekhar, C. S. C. (2004), 'Agriculture Price Volatility in International and Indian Markets', *Economic and Political Weekly*, Vol. 39, No. 43, 23 October, pp. 4729–36.

Shreyas (2007), *Increasing Suicides in Wayanad: A Study Report*, Sultan Bathery, Wayanad, Kerala: Shreyas.

Subrahmanian, K. K. (2007), 'Impact of the WTO on Plantation Crops of South India: Export Performance and Price Instability', mimeo, Thiruvananthapuram: Centre for Development Studies.

Vaidyanathan, A. (2006), 'Farmers' Suicides and the Agrarian Crisis', *Economic and Political Weekly*, Vol. 41, No. 38, 23 September, pp. 4009–13.

10

Agrarian Crisis in Punjab
High Indebtedness, Low Returns, and Farmers' Suicides[1]

KARAM SINGH

Agrarian crisis is not the making of a single day. This is more so for Punjab, which had been in the forefront of Green Revolution technology and had high growth rate for a long period (Singh 2003; Kalkat et al. 2006). There were many sources of high growth, which acted to raise productivity per unit of land and production per year. As agriculture progressed in Punjab, the limits of these sources of growth in farm incomes was reached, and thus the growth in farm incomes slowed down. The state entered a phase where farming was progressing but the farmers were becoming poorer (Punjab Agricultural University ([PAU] 1998). As capital intensification of agriculture takes place, the cash costs assume significant proportion and the cost of capital investment replacements also increases, which act as a drain on farmers' incomes. In the process of growth, the pricing parity, if favourable, might provide some reprieve, but otherwise would worsen the farmers' plight. The history of economic development elsewhere points to people shifting out of agriculture but in India as well as in Punjab, the agriculture sector continues to be burdened with the same proportion, and in absolute terms even more pressure, which further lowers down the per capita incomes of the farmers. Agricultural credit from formal institutions plays an important role in the growth process, but may not keep pace with the growing demand, forcing farmers to resort to high cost informal sources for meeting the deficit needs of capital costs and consumption needs. The economic progress also entails continuous improvements in the living standards and extra expenditure on social festivities, which becomes difficult to curtail, thus leading to more distress and crisis.

The second section of the chapter traces the trends in growth rates and the emerging stagnation in agricultural incomes. The third section presents trends in agricultural costs and prices that account for declining farm income. The fourth part deals with the indebtedness of Punjab farmers and shows that some regions and some farmers have suffered the most. The fourth section is an account of the factors leading to farmers' suicides. The policy conclusions are presented in the last part.

GROWTH OF PROGRESSIVE AGRICULTURE IN PUNJAB

Punjab played a pioneering role in ushering an agriculture-led growth that paid rich dividends to its peasantry. The Green Revolution technology found its push in Punjab since the mid–1960s, initially in wheat, and subsequently in rice, and the rice–wheat cropping system rapidly came to dominate the state's agriculture. With about 3 per cent of India's net sown area and 1.5 per cent of its farming population, Punjab for a long time, year-after-year, produced about 10 per cent of the country's rice and 20 per cent of its wheat. The 1970s and 1980s were the heydays for the farmers of Punjab and the farming as productivity of both crops increased rapidly, prices were assured, marketing was hassle free, farm incomes rose, and the state led the Indian agriculture scene.[2]

The consolidation of holdings completed by the early 1960s acted as catalytic for farmers for making land improvements and investments in irrigation, which got a further boost with rural electrification, providing electricity to every village by the early 1970s. The crash programme for rural roads was started in 1968, and in three years, 4344 kms of rural roads were added, which was more than what was added till then (4154 kms) since 1950–1. Further, more than 10,000 kms were added during the next three years. By the early 1970s, all villages were also linked with metalled roads, which gave a boost to communication and development of markets. Since 1968, additional market yards have been set up for wheat and rice procurement so that no farmer has to travel for more than five kms for the sale of his major produce. As of today, there are 144 principal markets, 530 sub-yards, and more than 900 purchase centres (seasonal) for the purchase of rice and wheat. The Minimum Support Price (MSP) of wheat and rice has been effectively implemented in Punjab, and hardly any sale by any farmer has ever been reported to be below MSP for these crops.

Institutional credit to Punjab farmers for meeting investment needs such as irrigation, land development, and machinery, and for working capital needs played a very significant role. There are about 4000 primary agricultural cooperative societies with a membership of more than 21 lakh, 87 State Agricultural Development Bank branches with more than 8 lakh members, and more than 1000 bank branches of Scheduled Commercial Banks with more than 7 lakh farmers' accounts. Together, these sources had outstanding advances of more than Rs 12,000 crore to farmers as of 31 March 2006.[3] Punjab has the largest proportion of irrigated area, highest cropping intensity, and the most intensive use of chemical fertilizers and other inputs.[4]

Deceleration in Agricultural Growth

Agricultural growth in Punjab has decelerated from 4.6 per cent in the 1980s to 2.5 per cent in the 1990s and 1.9 per cent in the 2000s. The corresponding national average is 3.2 per cent for the 1990s. This deceleration is because of the dramatic slowdown in the crop sector, from a growth rate of 4.0 per cent in the 1980s to 1.1 per cent in the 1990s, with little improvement, if any, in the 2000s. The productivity gains in the rice–wheat cropping system have slowed down, reaching a plateau. And with cotton as the third most important crop facing severe pest attacks during 1996 to 2002, which led to almost a failure and showed negative yield growth; the Bt cotton varieties, picking up only recently do have promise but is also not problem free. The growth in the livestock sector though was significant at 5.8 per cent in the 1980s and 4.9 per cent in the 1990s, but has also come down to 2.5 per cent in the 2000s (Table 10.1).

There has been a systemic decline in the growth rate since the 1980s. The decadal growth rate of the crop sector was above 4 per cent for the decade 1980–1 to 1990–1.[5] It came down to just about 2 per cent for 1984–5 to 1994–5. It has been less than 2 per cent for the decades beginning 1985–6. The livestock sector, which has been a saviour for Punjab farmers, also had a peak growth rate for 1983–4 to

Table 10.1

Decadal Growth Rates of NSDP at Constant Prices: 1980–1 to 2004–5

Period	Agriculture and livestock	Agriculture	Livestock	Primary	Secondary	Tertiary	Total
1980–1 to 1990–1	4.497	4.035	5.848	4.644	8.708	4.964	5.384
1984–5 to 1994–5	3.135	2.088	6.082	3.556	7.904	4.249	4.511
1990–1 to 2000–1	2.314	1.123	4.943	2.496	6.785	6.062	4.562
1994–5 to 2004–5	2.071	1.408	3.414	2.193	4.753	6.669	4.322
2000/1 to 2004/5	1.708	1.305	2.473	1.891	4.562	6.562	4.230

Source: Government of Punjab, Statistical Abstract, various issues.

1993–4 at 6.1 per cent. It also declined since then and came down to about 3.4 per cent in the last decade.

Thus, the first signal of decline in the growth rate of farm incomes has been set since the early 1990s. By 1990–1, the major sources of growth, viz. land productivity (yield per ha), land use (cropping) intensity (production or income per year), and the crop substitution (high and stable income crops for low-income crops) reached close to the potentials. Since then, the growth rate in yield has been either very low and declining (wheat) or stagnant (rice). There were major crop failures, severe in some years (cotton).[6] The index number of yield, with 1970–1 as the base, was 138 in 1980–1, 187 in 1990–1 and has been fluctuating between 200 and 215 since 2000–1. There has been no increase in the cropping intensity since 1997–8, which was 140 in 1980–1, 176 in 1990–1 and 187 in 1997–8. And the growth in other high-income and high-value alternative crops remained almost negligible, constrained by the limited market support.

FARM COSTS, PRICES, AND PROFITABILITY

Punjab has achieved very high levels of productivity per year through increase in cropping intensity and intensive use of inputs. At high levels of output, the exploitation of natural resources (soil nutrients and micronutrients) per unit of output is even higher. It led to micronutrient deficiencies, namely zinc, sulphur, iron, etc. in many areas in the state. Therefore, in order to maintain and improve the yield, the farmers have been further increasing the use of inputs. Fertilizer use (nitrogen, phosphorous, potassium [NPK] nutrients) per hectare of net sown area increased from 273 kg in 1990–1 to 309 kg in 2000–1 and further to 371 kg in 2004–5. This is an increase of 13 per cent during the 1990s and 20 per cent during the first four years of the 2000s.

The increase in the prices of inputs has been even higher. The price of urea increased by 50 per cent during the 1990s (from Rs 3060 per tonne in 1991 to Rs 4600 per tonne in 2002) and that of Di-ammonium Phosphate (DAP) by 90 per cent (from Rs 4680 per tonne to Rs 8900). The price of urea and DAP further increased to Rs 4830 and Rs 9350 per tonne by 2005. The price of weedicides, which also became an important input, increased by three times from Rs 160 per kg to Rs 460 per kg and that of diesel oil by more than four times from Rs 4.07 per litre to Rs 17.50 per litre between 1991 and 2002. In the next three years, by 2005, there were further increases in the prices—of weedicides to Rs 558 (22 per cent) and diesel to Rs 30.26 (73 per cent).

The total operational costs of rice and wheat increased by more than 50 per cent in 2005–6 over 2000–1 (Table 10.2). But the yield of rice increased by only about 12 per cent in five years and that of wheat even declined by about 8 per cent during the same period. From 1990–1 to 1995–6, the gross margin from rice improved by 102 per cent, and from 1995–6 to 2000–1 it increased by another 77 per cent. But since 2000–1, the increase has been only by 7 per cent in 2005–6 over 2000–1. The MSP for paddy during 1990–1 was Rs 205, it was raised to Rs 360 in 1995–6 and to Rs 530 in 2001–2. In response to improved price and gross margin from rice, the area under rice further increased by 15 lakh acres between 1990–1 and 2000–1. The investment in tractors went up and the number of tractors has increased from 275,000 to 405,000 during the same period. In the case of wheat, which is an even more stable crop and covers a larger area than paddy, the gross margin even declined by 34 per cent during 2000–1 to 2005–6. This has been on account of

Table 10.2
Changes in Costs and Returns of Rice and Wheat in Punjab:
1995–6 to 2005–6

(Percentages)

Item	Rice		Wheat	
	1995–6 to 2000–1	2000–1 to 2005–6	1995–6 to 2000–1	2000–1 to 2005–6
Operational cost	40	49	39	53
Cash cost	45	53	46	62
Total cost	44	49	58	66
Yield	12	12	18	8
Price	42	12	69	7
Gross income	59	26	99	−1
Gross margin	77	7	67	−34
MSP	42	12	69	7

Source: Derived from the Cost of Cultivation data of the Commission for Agricultural Costs and Prices, Government of India. (The data are collected by Punjab Agricultural University from 30 village clusters from 300 farmers on cost accounting basis.)

almost a freeze in the MSP of wheat and paddy during the last five years. Compared with 42 per cent and 69 per cent increase in MSP of paddy and wheat during 1995–6 to 2000–1, over the next five years the increase was only 12 per cent and 7 per cent respectively.

Relative Cost of Agricultural Capital Equipment

Punjab agriculture has become highly capital-intensive. And the common indicator often reported is that of number of tractors, whose number has increased from about 25,000 in 1970 to 405,000 in 2005. Punjab is considered to have too many tractors. However, in the farmers' perception, tractors improve timeliness of operations and improve yield, besides ease of operations and other associated works accomplished. Even technically, about 0.4 million tractors for 4.0 million hectares of cultivated area means a tractor for every 10 hectares, which is about the farm size that a tractor would handle technically efficiently (Singh et al. 2004). This means no surplus. But, the concern is how the farmers face the parity-price of agricultural capital formation vis-à-vis the prices of their produce. In 1970, a

Table 10.3
Parity of Capital Investments in Punjab: Tractor versus Wheat/Car

Year	No. of tractors ('000)	Price of tractor (Rs)	Price of wheat (Rs per tonne)	Tonnes of wheat to buy a tractor	Tractor/ Wheat parity	Price of car (Zen) (Rs)	Tractor/ Car Parity
1970	25	21,873	760	28.8	100		
1980	110	62,500					
1990	265	142,685	2,250				
1995	330	216,600	3,600	60.2	48	338,000	100
2000	410	285,000	4,450				
2005	440	340,000	5,600	60.7	47	340,000	64

Source: Calculated from the average prices prevailing.

tractor was costing about Rs 22,000 which could be sourced by selling 28.8 tonnes of wheat (Table 10.3). In 2005, a tractor cost Rs 340,000 for which the farmer had to sell as much as 60.7 tonnes of wheat, which is more than double the quantity of 1970. Thus, the parity of capital investments in agriculture has reduced to half in three decades. Looking through another barometer, in 1995 the cost of a car (Maruti Zen) was Rs 338,000 and that of a tractor was Rs 216,600, which means that one could buy three tractors in the price of two cars. In 2005, the price of a tractor is the same as that of a car, that is, Rs 340,000. Thus the parity in relation to the industrial goods has come down to 64 per cent in one decade. The agricultural terms of trade analysis, whichever the method, does not bring this message of growing disparity as effectively as this simple direct analysis.

Replacement Cost of Capital

The machine expenditure in relation to farm size has undergone a change over time—from positive association in the 1980s to a negative one in the 1990s onwards. The small farmers now have to spend even more as machine expenditure and much more so as cash costs because they hire in much larger proportion than other farmers who own the machinery. Strikingly, the hiring-in cost of farm machinery constituted more than 60 per cent of the operational machine expenses on small farms and less than 30 per cent on large farms. Capital costs have become much more, and the cost of replacements is increasing even faster (Singh 1990; Singh & Kalra 2002).

Add to these the capital investments required in deepening the tubewells, which was conservatively estimated at Rs 160 crore at 1999 prices and increasing in real terms at the rate of 7.2 per cent per year (Singh & Kalra 2002). This does not include the cost of going for submersible pumps, which by now have become about 25 per cent of the total tubewells in the state. This cost is much higher but the farmers still have no worthwhile option. And this is going to remain a big drain on farmers' incomes in the conceivable future.

Punjab faces a water and electricity crisis that deepens year by year. The water table in central Punjab, which has sweet underground water, had gone down in height by more than three metres till 1997 over 1973, and at an alarming rate of 76 centimetres per year between 1997 and 2006 (Singh 2006). As many as 66 per cent of blocks (80 per cent of sweet water blocks) have gone dark, that is, where the groundwater withdrawal is more than 85 per cent of available recharge and the water table has gone too deep (Hira 2006; Hira, Jalota & Arora, 2004; Kalkat et al. 2006). The number of tubewells with submersible pumps in 2005 went up to about 300,000. It costs about Rs 80,000 to Rs 100,000 for a submersible pump installation. And their number is increasing every year as more shallow tube wells are failing. The estimated cost up to 2010 has been placed at Rs 4000 crore (Kalkat et al. 2006; Singh 2006). And this is an added source of drain on farmers' incomes in recent times. Thus, the farmers' indebtedness goes on increasing. Overcapitalization of the Punjab farmers and increasing cost of capital replacements have become a big financial burden.

Declining Profitability

The recent trend in overall profitability in terms of value of output per rupee of input has been on the decline. An analysis of the cost of cultivation of wheat, the major crop in the state, brings this out (Table 10.4). There was a 15 per cent decline in profitability (value of output per unit of input, that is, all costs) from 1.47 in 1999–2000 to 1.25 in 2003–4. The fall in profitability in respect of operational costs was even higher at 18 per cent, from 3.12 in 1999–2000 to 2.55 in 2003–4. It may be noted that a 15 per cent decline in profitability requires a 15 per cent increase in the volume of business to keep one's income even

Table 10.4

Trends in Profitability of Wheat in Punjab: 1999–2000 to 2003–4

Year	Total VOP (Rs per ha)	Cost C2 (TC) (Rs per ha)	Operational Costs (OC) (Rs per ha)	Cost C2 (Rs per qtl)	Net margin (Rs per ha)	Profitability Ratios		Price Index	Real Net Margin Rs per ha in 1999–2000 prices
						VOP/TC	VOP/OC		
1999–2000	31,247	21,312	10,000	396	9,935	1.47	3.12	100	9,935
2000–1	31,804	22,537	10,382	432	9,267	1.41	3.06	104	8,911
2001–2	31,172	22,931	11,045	456	8,241	1.36	2.82	108	7,631
2002–3	29,200	22,997	11,653	494	6,203	1.27	2.51	113	5,489
2003–4	28,033	22,415	10,978	504	5,618	1.25	2.55	117	4,545
% change	–10.3	5.2	9.8	27.3	–43.5	–15.0	–18.2		
2006–7	35,768	27,043	15,857	943	8725	1.32	2.26	134	6,513

Note: Cost C2 includes all actual expenses in cash and kind incurred in cultivation, rent paid for leased-in land and imputed value of family labour, interest in value of owned capital assets (excluding land), and rental value of land (net of land revenue), Ha denotes hectare, Qtl denotes quintal, Rs denotes Rupees, TC and OC denote total cost and operational cost respectively, VOP denotes value of produce.
Source: Cost of Cultivation Data. Reports of the Commission for Agricultural Costs and Prices, Government of India. For the relevant years.

at a stagnant level. But a decrease in the volume of business means an impending crisis. And a crisis indeed it became as the value of output (at current prices) declined by 10.3 per cent over this period. The rising costs squeezed the farmers' net margins further; from about Rs 10,000 per ha in 1999–2000 to about Rs 5600 per ha in 2003–4, that is, a crunch of 43 per cent at current prices in five years. The cost of living (gauged through the Consumer Price Index) during this period increased by 17 per cent; thus the real income of an average wheat farmer declined to less than half in just five years.

AGRICULTURAL CREDIT AND FARMERS' INDEBTEDNESS IN PUNJAB

It is the general belief that institutional credit through a widespread network of cooperatives and commercial banks complemented Punjab's agriculture sector in achieving many landmarks through intensive input use and a very high growth of capital investments on farms (Singh 1990). Even in Punjab, although the bank lending did not show a decline, the number of direct agricultural bank accounts declined. In 1990, there were over 7 lakh bank accounts of direct finance to the Punjab farmers, which steadily declined during the 1990s and reached close to 5 lakh by 1999. There has been an increase in these accounts since then, and in 2003 there were 6.5 lakh accounts of direct finance to the farmers. The fact is that the bank lending is still inadequate, costly and cumbersome, particularly for the small and marginal farmers. Therefore, the farmers have to resort to the private non-institutional sources of finance, which have their own ways of exploitation and squeezing the farmers' net incomes.[7] There was decline in rural (in settlements with less than 10,000 population) bank branches in Punjab, from 1178 in 1990 to 1111 in 1996, a marginal increase to 1126 by 2003. There are 12,278 villages in Punjab, which means that less than 10 per cent villages have a bank branch. There is one bank branch per 11 villages in Punjab.

The direct finance to farmers by the commercial banks increased from Rs 235 crore in 1980 to Rs 1070 crore in 1990 and further to Rs 5567 crore in 2004. However, seen as a share of Net State Domestic Product (NSDP) from agriculture, the credit situation deteriorated since 1990. It was about 7 per cent of NSDP from agriculture in 1980 and increased to about 16 per cent in 1990. It started declining since 1991 and came down to about 10 per cent of NSDP from agriculture during 1994–2000. But, in recent years, commercial bank credit as a per cent of NSDP from agriculture increased from less than 11 per cent in 2000, to more than 22 per cent in 2006. This increase is partly due to the growing realization of the need to rescue agriculture and partly due to the contracting agriculture income.

In respect of institutional credit, the cooperatives in Punjab have played a key role. There are 4000 Primary Agricultural Cooperative Credit Societies (PACS) in Punjab, which continue to dominate in the delivery of short-term credit to the farmers. The outstandings to PACS increased from Rs 482 crore in 1990–1 to Rs 2665 crore in 2004–5, which is a growth rate of 12.99 per cent. The short-term credit, especially for fertilizers to the farmers at their door steps, has remained the major domain of the PACS in Punjab. Their recovery has been far too good and they continue to play their assigned role successfully. Another arm of the cooperatives, the State Agricultural Development Banks (SADBs) have looked after the term credit. The outstandings to the SADBs also increased from Rs 311 crore in 1990–1 to Rs 2012 crore in 2003–4, which is an even higher growth rate of 14.27 per cent. The recovery performance of the SADBs has been even better in Punjab and accordingly, the overdues were well below 10 per cent for most of the years except for 2003–4 and 2004–5, since when the agrarian crisis has surfaced.

The outstanding advances from the institutional sources alone increased from Rs 1863 crore in 1990–1 to Rs 5377 crore, in 1998–9 and further to Rs 8188 crore in 2002–3, and to Rs 12,411 crore in 2005–6. As per cent of NSDP from agriculture, it increased from 19 per cent in 1996–7 to 23 per cent in 1998–9. Since then, it has increased much faster, to 25 per cent of NSDP from agriculture in 2000–1, 33 per cent in 2002–3, and to 38 per cent in 2005–6 (Table 10.5).

The institutional credit to the farmers has increased over time but it has not been adequate enough to make a really significant dent on the non-institutional lending to the farmers. The institutional credit to the farmers also comes at a cost, other than the rate of interest. It is fraught with many inadequacies such as amount, easiness, timeliness, and purpose with many strings of formalities required for which the farmers have to spend extra money, including bribes.[8] For this reason, the farmers still have to approach the non-institutional sources for their credit requirements.

Table 10.5

Outstanding Advances from Institutional Sources to Punjab Farmers

(Rs crore)

Year	Commercial banks	Primary Agricultural Cooperative Societies	State Agricultural Development Banks	Total	Incremental	NSDP from agriculture and livestock	Total institutional debt as per cent of NSDP
1970–1	20#	52	48	120		834	14.4
1980–1	235	146	109	490		2,023	24.2
1990–1	1,070	482	311	1,863		7,393	25.2
1991–2	1,167	512	342	2,021	158	9,423	21.4
1992–3	1,249	560	374	2,183	162	10,916	20.0
1993–4	1,332	605	396	2,333	150	12,962	18.0
1994–5	1,472	665	472	2,609	276	14,264	18.3
1995–6	1,655	756	551	2,962	353	15,369	19.3
1996–7	1,823	832	740	3,395	433	17,927	18.9
1997–8	1,982	1,043	957	3,982	487	18,750	21.2
1998–9	2,230	1,327	1,240	4,797	815	20,544	23.3
1999–2000	2,528	1,410	1,439	5,377	480	23,137	23.2
2000–1	3,120	1,530	1,556	6,206	829	24,716	25.1
2001–2	3,476	1,740	1,663	6,879	673	25,622	26.8
2002–3	4,425	2,084	1,679	8,188	1,309	24,908	32.9
2003–4	5,567	2,319	1,769	9,655	1,467	27,333	35.3
2004–5	6,396*	2,665	2,012	11,073*	1,418	29,408	37.6
2005–6	7,225*	3,080*	2,106*	12,411*	1,338	32,347*	38.4

Note: #Derived from the 1975 figure of direct advances of Rs 27 crore. *Provisional
Source: Government of Punjab, Statistical Abstract, various years.

Over the years, the non-institutional sources of farmers' indebtedness have been increasing. According to the All-India Debt and Investment Surveys, the share of non-institutional sources of debt of farmers in Punjab increased from 18.1 per cent in 1991–2 (RBI 1999) to 52.1 per cent in 2003 (NSSO 2005). The latter result is similar to the one reported by a survey sponsored by the Punjab Government in 1997. In another survey in 2003, the share of non-institutional debt of farmers is shown as 57.7 per cent (PAU 2003).

Indebtedness of Punjab Farmers

In 1997, a study based on field survey estimated that the farmers' total debt was to the tune of Rs 5700 crore (Shergill 1998). The recent estimate for 2003 by the NSSO puts the outstanding debt of Punjab farmers at Rs 4574 crore (NSSO 2005). But an independent study for the same year puts the estimated indebtedness of farmers at a much higher level (Singh and Toor 2005). The variation in different estimates of indebtedness of the Punjab farmers notwithstanding, the overarching conclusion is that Punjab farmers are heavily indebted to the tune of more than 50 per cent of the NSDP from agriculture.

The costs of funds borrowed from the institutions vary according to the purpose for which these are borrowed. For instance, the cost of loan for the cash requirements for the crops, such as for buying fertilizers, does not require any payments for crop insurance or any other premium. The loan for dairying, however, requires the premium to be paid for insuring the dairy animal. Thus, the increase in loans for dairy, or for poultry, place a higher burden on the farmers than some other loans. The inadequacy of institutional loans and the farmers' requirements for purposes for which institutional loans are either not available or are costlier or even difficult to obtain increases the farmers' dependence on the non-institutional sources of funds, and hence the higher burden of interest payments.

There are no direct estimates of the total interest burden of the farmers. An attempt is made to estimate these indirectly from the cost of cultivation data for 2000–1 and projected to the present.[9] The cost of cultivation includes the interest as a cost separately for the total working expenses and fixed cost; it also includes the interest on the owned capital invested by the farmer from his own sources. The cost of cultivation data (2000–1) gives the interest cost per hectare at Rs 1418 for paddy and Rs 1940 for wheat, that is, Rs 3358 per hectare for two crops in one year, and for 4.2 million hectares of net area sown in Punjab the interest cost of Punjab, farmers for 2000–1 works out to Rs 1400 crore. The main reason for the high interest cost is the growing dependence on high interest-bearing non-institutional sources.

It is important to cut down the farmers' burden of interest costs through pumping in more of the institutional loans, which carry a lower rate of interest. But it should also be borne in mind that a lower rate of interest is of no use if the loan is not adequate, not timely, and with the additional transaction costs. The demand for loan of the large farmers has been reported to be met by the formal sources up to 67–75 per cent; but in the case of all other categories, it has been only 10–50 per cent. Also, in the case of formal sources, 'the additional costs involve cost of frequent visits to the institution, payment of a fee, submission of documents (requiring payment to some "munshi" who can do the paper work for the illiterate farmers), etc. These expenses, if added to the rate of interest, make formal borrowings quite expensive' (Gill & Singh 2006). In the process, the formal credit might even become costlier than the informal credit.[10]

INDEBTEDNESS AND FARMERS' SUICIDES IN PUNJAB

The increase in indebtedness of the farmers has resulted in farmers committing suicides in Punjab. According to a survey in 2005 by the Government of Punjab, 2116 farmers have committed suicides in the state during the last 15 years. It has to be acknowledged that the suicide rates in Punjab have been increasing at a higher rate than the all-India rate. Between 1985 and 1990, the all-India average grew by 27.14 per cent, whereas Punjab showed an increase of 35.85 per cent. Similarly between 1990 and 1995, the all-India increase was 8.99 per cent, whereas in Punjab it increased by 136.11 per cent (Satish 2006). During 1975 to 2001, the age-adjusted Suicide Mortality Rate (SMR) increased by 2.4 per cent per annum for males in India and at 2.8 per cent per annum in Punjab. During 1995 to 1998, the years of crop (cotton) failure, the SMR for farmers in Punjab was higher than the general SMR for males, by more than 33 per cent (Mishra 2006).

Suicide is a complex phenomenon. Most of the suicides in the rural areas in Punjab are by the farmers and the majority of the farmers committing suicides are under debt, which they have not been able to repay. The crisis of crop failures, particularly in the cotton belt of the state, and the resultant indebtedness of the farmers became so severe that three villages of the state in the Bathinda district were declared by the Village Panchayats as 'villages for sale'. The Village Panchayats in a declaration wrote to the state government to acquire all the resources, including land of the villages, and asked for getting rid of credit (Singh and Toor 2005). 'Several years have passed and the sign board is still outside Harkishan Pura (Bathinda District) "Village for sale"' (Gill 2006).

Marginalization and Agriculture

The Punjab farm sector, like elsewhere in India, is also saddled with a large number of small and marginal farmers operating up to 2 hectares of land. They constitute more than 35 per cent of the operational holdings (total number of operational holding in Punjab is just around 1 million) and are operating only 8.7 per cent of the cultivated area (total cultivated area in Punjab is 4.2 million hectares). The marginal farmers are leasing out their land in distress, and without any quality alternative employment in rural areas, find it even harder to live (Kalkat et al. 2006).

The rural sector in Punjab is lagging far behind in development; the rural–urban disparity is widening. In 1970–1, the rural population constituted 76 per cent; the share of NSDP from agriculture and livestock was 58 per cent. Over time, the share of agriculture and livestock in NSDP has declined but the share of rural population in the total remained almost the same. In 2004–5, the rural population constituted 66 per cent but the share of agriculture and livestock in the NSDP declined to 37 per cent. Worse still is the fact that the gross capital formation in agriculture and livestock sector has been a bare minimum at around eight per cent of its NSDP. Like education and health, agriculture has been a soft target for budget cuts. A very low proportion of the budget (3 per cent) is being spent on agriculture and allied activities in Punjab (Gill 2006). The marginalization of farms is accompanied by marginalization of agriculture in the priorities of budgetary provision. The political economy of the marginalization process operates in several dimensions. The rising costs, especially the growing interest burden, are already discussed above. The interest burden is disproportionately more on small and marginal farmers, whose borrowings are relatively more from the non-institutional sources and who also have very little of their own capital investments. They remain more entrapped with the high interest payments only.

A recent (2006) large-scale survey at the behest of the Punjab State Farmers' Commission (PSFC) shows that 89 per cent farmers were indebted. There were multiple sources of credit and 86 per cent of the borrowings were from the institutional agencies, which provided 66 per cent of the total loans. As many as 48 per cent farmers borrowed from the non-institutional agencies, which accounted for 34 per cent of the total loans to the farmers.[10] Since 2005–6, on account of a spurt in the farmer suicides and consequently the government agencies becoming more active, the non-institutional agencies have become somewhat cautious and restrictive. Of the three regions of the state, that is, the hilly region (eastern zone, 9 per cent area), the rice region (central zone, 64 per cent area), and the cotton region (western districts, 27 per cent area), the cotton region is the most disadvantaged. The average indebtedness in the cotton region is Rs 64,999 per hectare compared to Rs 43,275 in the rice region and Rs 18,177 in the hilly region (PSFC 2007). On a per hectare basis, the marginal farmers are almost three times (Rs 101,321 per ha) more indebted compared to large farmers

Table 10.6
Farm Size and Indebtedness in Punjab: 2006

Farm size class/Region	Average farm size (Ha)	Average debt per ha (Rs)	Share of non-institutional sources (Rs per ha)	Share of non-institutional sources (%)
Hilly region	2.08	18,177	6,689	36.8
Rice region	3.52	43,275	12,030	27.8
Cotton region	4.10	64,999	27,624	42.5
< 1 ha	0.71	101,321	34,347	33.7
1 –2 ha	1.64	68,549	23,443	35.3
2 –4 ha	3.10	67,807	23,800	35.1
4 –6 ha	5.08	42,332	15,917	37.6
> 6 ha	7.32	35,363	7,744	21.8
All	3.57	50,140	17,699	35.3

Source: PSFC (2007).

owning more than six hectares (Rs 35,363 per ha). Except large farmers, all other classes of farmers depend to the extent of one-third on the non-institutional sources (Table 10.6).

What emerges from these studies is that the cotton region and the small and marginal farmers are more prone to suicides, and that is what has been reported. Although there are other reasons such as family discord, alcoholism and drug abuse, loss of status, and lack of resources, it is the indebtedness, especially the inability to repay, that acts as the major cause for farmers' suicides. Table 10.7 provides an abstract of reasons captured by major studies on suicides in Punjab. In addition, one can also refer to Sidhu and Gill (2006), Satish (2006), and Gill and Singh (2006).

The resource-poor farmers reporting suicides constitute the largest proportion of suicide victims. Indebtedness, more significantly, borrowing from more than one source, and particularly the private

Table 10.7
Profile of Suicide Victims in Punjab

Item	Bhalla et al. 1998	Iyer and Manick 2000	AFDR 2000	PAU 2003	Chahal 2005
Suicide cases studied	53	75	79	30	42
% cultivators	55	67	85	100	100
% agricultural labour	45	33	15		
% small and marginal farmers	25	84	66		55
Causes of suicides amongst suicide cases (multiple responses—% reporting):					
i. Indebtedness	38	79	62	87	55
ii. Crop failure	1	10	5	3	7
iii. Family discord	36		6	27	55
iv. Loss of status	17			10	
v. Lack of resources	6		60		
Debt exclusively from moneylenders	37	68	27		9
Debt from moneylenders and other sources	n.a.	81	74	100	52
Unproductive use of loans	68	52	20		39

Sources: Bhalla et al. (1998), Iyer and Manick (2000), AFDR (2000), PAU (2003), Chahal (2005).

sources, account for the maximum proportion of farmers' suicides. The unproductive use of the loan, which is more due to social stress than choice, is also a factor responsible for a large proportion of suicides in Punjab.

Indebtedness is the main but not the lone cause of suicides. This is captured by the PAU study, which showed that 87 per cent suicide cases had indebtedness as the main trigger force, which is aggravated by other forces. Out of these, 20 per cent are pure credit related (harassment from moneylenders), 20 per cent are income decline related (crop failure: 10 per cent and status loss: 10 per cent), 27 per cent are social discord related (sudden expenses, family discord 13 per cent and family responsibilities 14 per cent) and 20 per cent others (mental trauma 17 per cent, drug/alcohol abuse 3 per cent). Thus, social evils such as dowry and the extravagant expenditure on social ceremonies, particularly by the lower income strata of population, aggravate the problem. The resource-poor farmers' suicide indicates that there is a breakdown of community feeling and social support mechanism in areas of highly commercialized and competitive agriculture (Sidhu & Gill 2006).

SUMMARY AND POLICY IMPLICATIONS

Agricultural growth in Punjab has slowed down. The cost of inputs has increased faster than the output prices, especially during the last more than ten years, beginning with the mid–1990s. The cost of capital investments has increased faster in the farm sector; with the farming having become much more capital-intensive in Punjab, the cost of capital investment replacements has increased manifold and consumes a significant proportion of the farmers' incomes. The drag on natural resources, especially groundwater, has further dragged the farm incomes towards digging deeper for submersible pumps installation. Profitability declined substantially and Punjab farmers, who were used to a higher standard of living during the era of high growth in their incomes, now suffer income squeeze and often try to keep up their lifestyle by increased borrowing.

Credit did play an important role in the growth of the agricultural economy in Punjab. While institutional lending to the farmers increased, it remained far short of their demand. The non-institutional agencies, mainly the commission agents and the moneylenders, have come to provide an increasing share of credit. The result is growing interest burden of farmers. The small farmers with limited access to institutional credit are forced to borrow relatively more from the non-institutional agencies, and also at times, rotate credit from non-institutional to institutional sources, and vice versa, leading to exploitation by multiple agencies. In the cotton region, the non-institutional credit is far more than in the rice and hilly regions of the state.

The marginal and small farmers get entrapped in tractor loans, where bankability needs more careful assessment; the unorganized and unregulated second-hand tractor markets with farmers as the weaker party have become a common source of distress sales in Punjab. Even at the end of the second normal year in March 2006, about 19 per cent of the marginal and small farmers were under acute burden of indebtedness, that is, more than 200 per cent of their annual income compared with less than six per cent of the medium and large farmers. The marginal, small, and semi-medium farmers with holdings up to four hectares were indebted up to about 90 per cent of their income; the medium and large farmers up to 50 per cent of their annual income; and in the cotton region it was 91 per cent.

The agrarian crisis requires a comprehensive strategy that encompasses not only immediate relief to suicides but also revival of agricultural growth and assured incomes to the farming community. Public investment in agriculture should be increased up to about 12.5 per cent of NSDP on a long-term basis. Public investment should include marketing infrastructure that facilitates farmers' organizations to take up the processing and marketing, which will help improve the farmers' incomes. The rural financial institutional network must be strengthened. The formal sector loans have to be adequate, timely, more economical, and commensurate with demand. The extra costs mandatory on the loans such as insurance of the tractor or the dairy animals, should be met out of some consolidated fund to be created by the government. The financial institutions should also provide incentives to promote a savings culture among farmers. There is need for credit counselling to farmers.

The rural cooperatives should be enabled to go beyond mere provision of credit. If rural cooperatives are enabled to acquire and lease agricultural machinery, such as tractors which are needed for a short period in a season, most of the small and marginal farmers can reduce their debt and interest burden substantially. There is need for safety net programmes for rural masses in general and farmers in particular, such as old age pensions and health insurance. On the recommendation of the Punjab State Farmers' Commission, the Government of Punjab launched the health insurance scheme called *Sanjivni* in April 2006. Any cooperative member aged up to 75 years can join the scheme by paying a fixed annual premium of Rs 300 and avail medical treatment costing up to Rs 2 lakh a year from 150 government and private hospitals across the state without paying any cash.[11] Primary members' dependents have to pay a nominal add-on premium of Rs 30 to get a similar health cover (Vinayak 2006). It is the perfect example of public–private partnership in making secondary and tertiary health care accessible to the rural masses. There is need for promoting rural non-farm enterprises not only to enable farm households to earn supplementary income but also to wean those on unviable farms gradually to shift to alternative better productive employment, which needs training in better skills.

Notes

1. The views are of the authors only.

2. Most of the data in the chapter comes from various issues of the 'Statistical Abstracts of Punjab', published annually by the Government of Punjab. Other sources, wherever used, are specifically mentioned.

3. This translates to Rs 120,000 per farmer (There are about one million farming households in Punjab) from the institutional sources alone in 2006. Compare it with the Rs 41,576 per farming household from both the institutional and non-institutional sources (which was also the highest amongst the states in India) brought out by the Situation Assessment Survey for 2003 (NSSO 2005). These issues are covered in more detail in the third section.

4. Punjab has 98 per cent irrigated area, 187 per cent cropping intensity, and 185 kgs of fertilizer nutrients use per cropped hectare compared with all India figures of 40, 132, and 90, respectively (GoI 2005).

5. The agriculture sector is synonymous with the crop sector; it includes all the general as well as horticultural (fruits and vegetables) crops, though the latter is still very insignificant. Then there is the livestock sector. These two along with forestry and logging, fisheries, and mining and quarrying constitute the primary sector .

6. Wheat, rice, and cotton occupied 80 per cent of total cropped area in Punjab in 1990–1 and 84 per cent in 2004–5.

7. Although the private moneylending is always/generally considered to be more expensive than formal sector loans, there could be instances when this may not be the case. On a recent field trip, a farmer's example is illustrative. He wanted to replace the old tractor, selling it and buying a new one. He was short of Rs.1 lakh. He approached the bank branch, and was insisted on: (i) borrowing the full amount, (ii) he could return back the surplus on the next day, (iii) the price he would get for his old tractor will not be discounted, and (iv) the tractor will have to be insured. He borrowed from the commission agent at the rate of 2 per cent per month; got a discount of Rs 12,000 on the tractor price and paid no insurance cost. He repaid the whole loan in six months; the interest cost came to Rs 12,000 only; the farmer saved the whole bank interest, other bank costs and the insurance premium for the tractor.

8. A recent field study by Punjab State Farmers' Commission (PSFC) of 600 farmers in 20 Blocks in Punjab showed that the transaction costs incurred by the farmers even before the bank loan is availed is about 5 per cent of the loan amount (PSFC 2007).

9. The latest year for which the data were available at the first count of this study. Although later on more recent estimates up to 2002–3 for rice and 2003–4 for wheat were available, it will not make much difference to the spirit of the estimates shown here. Five years is a good gap to simulate other economic forces into the system.

10. See notes 6 and 7.

11. Recently, under the supervision of Karam Singh, PSFC (2007) carried out a survey of 600 farmers. A preliminary estimate of the total institutional loans comes to Rs 12,998 crore, which is very close to the provisional figures given by the institutions.

12. The scheme has already covered nearly 5.73 lakh farmers and their families and is run under the aegis of a Punjab Government-patronized trust in conjunction with an insurance company, a third party insurance administrator, and select hospitals. During April–June 2006, 1800 members have availed medical treatment worth Rs 3 crore under the scheme and the maximum that any individual has availed of is Rs 1.86 lakh, a little short of the permissible upper limit of Rs 2 lakh.

REFERENCES

Association of Federal Democratic Rights (AFDR) (2000), *Suicides in Rural Areas of Punjab: A Report* (in Punjabi), Patiala: AFDR, District Unit, October.

Bhalla, G. S., S. L. Sharma, N. N. Wig, S. Mehta, and P. Kumar (1998), *Suicides in Rural Punjab*, Chandigarh: Institute for Development and Communication (IDC).

Chahal, T. S. (2005), *Forced Fall: A Case of Punjab Farmers*, Amritsar: Institute of Development and Planning.

EPW (2004), 'Doubling Rural Credit, But How?', *Economic and Political Weekly*, Vol, 39, No. 24, 12 June, pp. 2415–6.

Gill, A. and L. Singh (2006), 'Farmers' Suicides and Response of Public Policy: Evidence, Diagnosis and Alternatives from Punjab', *Economic and Political Weekly*, Vol. 41, No. 26, 30 June, pp 2762–8.

Gill, S. S. (2006), 'Plight of Punjab's Farmers: Agriculture Being Discriminated Against', Chandigarh, *The Tribune*, 22 August.

Government of India (1956), *All India Rural Credit Survey 1951*, Mumbai: Reserve Bank of India.

——— (2005), *Agricultural Statistics at a Glance*. New Delhi: Ministry of Agriculture, November.

Government of Punjab (various years), *Statistical Abstracts of Punjab*, Chandigarh: Government of Punjab.

Hira, G. S. (2006), 'Ground Water Management', Ludhiana: Department of Soils, Punjab Agricultural University (PAU).

Hira, G. S., S. K. Jalota, and V. K. Arora (2004), *Efficient Management of Water Resources for Sustainable Cropping in Punjab*, Ludhiana: Department of Soils, PAU.

Iyer, G. K. and M. S. Manick (2000), *Indebtedness, Impoverishment and Suicides in Rural Punjab*, New Delhi: India Publishers and Distributors.

Kahlon, A. S. and K. Singh (1984), *Managing Agricultural Finance: Theory and Practice*, New Delhi: Allied Publishers.

Kalkat, G. S., K. S. Pannu, K. Singh, and P. S. Rangi (2006), *Agricultural and Rural Development of Punjab: Transforming from Crisis to Growth*, Chandigarh: The Punjab State Farmers Commission (PSFC), Government of Punjab, May.

Mishra, S. (2006), 'Suicide Mortality Rates Across States of India, 1975–2001: A Statistical Note', *Economic and Political Weekly*, Vol. 41, No. 16, 22 April, pp. 1566–9.

National Sample Survey Organisation (NSSO) (2005), *Situation Assessment Survey of Farmers: Indebtedness of Farmer Households*, NSS 59th Round (January–December 2003), Report No. 498 (59/33/1), New Delhi: Ministry of Statistics and Programme Implementation, Government of India.

PAU (1998),'Proceedings of Brainstorming Meeting on Farmers and Farming in Punjab'. Ludhiana: PAU, October.

——— (2003), *Market Imperfections and Farmers' Distress* (D. K. Grover, Sanjay Kumar and Kamal Vatta), Ludhiana: Agro Economics Research Centre, PAU, September.

PSFC (2007), *Flow of Funds to Farmers and Indebtedness in Punjab,* Ludhiana: PAU and Chandigarh: PSFC.

Reserve Bank of India (1954), *All India Agricultural and Rural Credit Survey— Report of the Committee of Direction,* Vol. II, General Report (Chairman: A. D. Gorwala).

——— (1998, 1999),'All-India Debt and Investment Survey, 1991–92— Incidence of Indebtedness of Household', *RBI Bulletin,* Vol. 53, No. 5, May and Vol. 54, No. 2, February.

Satish, P. (2006), 'Institutional Credit, Indebtedness and Suicides in Punjab', *Economic and Political Weekly,* Vol. 41, No. 26, 30 June, pp. 2754–61.

Shergill, H. S. (1998), *Rural Credit and Indebtedness in Punjab,* Chandigarh: Institute for Development and Communication.

Sidhu, R. S. and S. S. Gill (2006), 'Agricultural Credit and Indebtedness in India: Some Issues', *Indian Journal of Agricultural Economics,* Vol. 61, No. 1, January–March, pp. 11–35.

Singh, K. (1990), 'Case Studies of Successful Farmers, Societies and Private Entrepreneurs: Experience and Issues', Keynote Paper, *Indian Journal of Agricultural Economics,* Vol. 45, No. 3, July–September, pp. 347 –54.

——— (2003), *Punjab Agricultural Policy Review,* New Delhi: World Bank, July.

——— (2006), *Fall in Water Table in Punjab: How Serious,* Chandigarh: PSFC, Government of Punjab, 22 November.

Singh, K. and S. Kalra (2002), 'Rice production in Punjab: Systems, Varietal Diversity, Growth and Sustainability', *Economic and Political Weekly,* Vol. 37, No. 30, 27 July, pp. 3139–48.

Singh, K. and P. S. Rangi (2006), *Second Hand Tractor Markets in Punjab: Need for Regulation* (An Appraisal), Chandigarh: PSFC, Government of Punjab, 1 November.

Singh, K., P. S. Rangi, and S. Kalra (2004),'Wheat Production and Sustainability in Punjab: Growth and Varietal Diversity', *Indian Journal of Agricultural Economics,* Vol. 59, No. 4, October–December, pp. 745–71.

Singh, S. and M. S. Toor (2005), *Magnitude and Determinants of Indebtedness in Punjab Agriculture,* Ludhiana: Department of Economics, PAU.

Tribune News Service (2006), 'Cabinet Panel for Waiving Loans of Farm Labour', *The Tribune*, 13 July.

Vinayak, R. (2006), 'Card to Rural Relief: A Novel Rural Health Insurance Scheme Helps Punjab Farmers Get Access to Expensive Treatment', *India Today*, 14 August, p. 9.

Contributors

Ramesh Chand	ICAR National Professor, National Centre for Agricultural Economics and Policy Research, Pusa, New Delhi.
R.S. Deshpande	Director, Institute for Social and Economic Change, Bangalore.
S. Galab	Professor, Centre for Economic and Social Studies, Hyderabad.
Vineetha Menon	Head, Department of Anthropology, Kannur University (Former Visiting Fellow, Kerala Research Project, Centre for Development Studies, Thiruvananthapuram).
Srijit Mishra	Associate Professor, Indira Gandhi Institute of Development Research, Mumbai.
K.N. Nair	Director, Centre for Development Studies, Thiruvananthapuram.
Suresh Pal	Senior Scientist, National Centre for Agricultural Economics and Policy Research, Pusa, New Delhi.
V.M. Rao	Visiting Honorary Fellow, Institute for Social and Economic Change, Bangalore.
D. Narasimha Reddy	Formerly Professor of Economics, University of Hyderabad, and Visiting Professor, Institute for Human Development, New Delhi.

P. Prudhvikar Reddy	Associate Professor, Centre for Economic and Social Studies, Hyderabad.
E. Revathi	Associate Professor, Centre for Economic and Social Studies, Hyderabad.
S.L. Shetty	Director, Economic and Political Weekly Research Foundation, Mumbai.
Karam Singh	Agricultural Economist, Punjab State Farmers' Commission, Government of Punjab (Former Director, Agro Economic Research Centre, Punjab Agricultural University, Ludhiana).